Only the bamboo knows the pure breeze

INSCRIPTION FROM AN INK PAINTING BY ZEN MASTER NANREI KOBORI-ROSHI, RYOKO TEMPLE, KYOTO, 1974

The Craft & Art of Bamboo

30 Elegant Projects to Make for Home and Garden

The Craft & Art of Bamboo

Carol Stangler

LARK BOOKS

A Division of Sterling Publishing Co., Inc.
New York

EDITOR: Janice Eaton Kilby
ART DIRECTOR: Dana Margaret Irwin
ASSISTANT ART DIRECTOR: Hannes Charen
ASSISTANT EDITOR: Heather Smith
EDITORIAL ASSISTANT: Rain Newcomb
EDITORIAL ASSISTANCE: Anne Wolff Hollyfield
PHOTOGRAPHY: Sandra Stambaugh

Library of Congress Cataloging-in-Publication Data
Stangler, Carol A., 1946-
    The Craft and Art of Bamboo: 30 Elegant Projects to Make for Home and Garden /
Carol A. Stangler.— 1st ed.
        p. cm.
    ISBN 1-57990-191-3
    1. Bamboo work. 2. Bamboo construction. 3. Bamboo furniture. I. Title

TT190 .S73 2001
745.51—dc21
2001029431

10 9 8 7 6 5 4 3 2 1
First Edition

Published by Lark Books, a division of
Sterling Publishing Co., Inc.
387 Park Avenue South
New York, N.Y. 10016

© 2001, Carol A. Stangler

Distributed in Canada by Sterling Publishing,
c/o Canadian Manda Group, One Atlantic Ave., Suite 105
Toronto, Ontario, Canada M6K 3E7

Distributed in Australia by Capricorn Link (Australia) Pty Ltd.,
P.O. Box 704, Windsor, NSW 2756 Australia

Distributed in the U.K. by Guild of Master Craftsman Publications Ltd.,
Castle Place 166 High Street, Lewes, East Sussex, England, BN7 1XU.
Tel: (+44) 1273 477374 • Fax: (+44) 1273 478606
Email: pubs@thegmcgroup.com • Web: www.gmcpublications.com

If you have questions or comments about this book, please contact:
Lark Books
50 College Street
Asheville, North Carolina 28801
(828) 253-0467

Printed in China.

ISBN 1-57990-191-3

Dedication

To my father, Bernard Stangler

whose career as nurseryman and landscaper
surrounded me with all things green and
growing, and whose passion for fishing and
the outdoors planted me firmly in the earth,

and to the memory of my mother, Inez Todnem Stangler

whose years as homemaker to our family of
nine gifted me with the art of creating order
and beauty, and whose lessons with needle
and thread form the basis of my work.

CONTENTS

BAMBOO

In the last 20 years, bamboo has literally taken hold in nooks and crannies throughout the Western world.

A few roots and stems called *cuims* planted here and there have grown into rambling, exotic groves, dotting the landscape with their magnificent, leafy forms. Growers have acquired and propagated bamboo plants suitable to a temperate climate, and Western land-scapers and gardeners have embraced bamboo for its lush foliage and striking form.

Likewise, bamboo poles, long used almost exclusively by Far Eastern and South American cultures, are now readily accessible. As Westerners realize the strength and resiliency of this giant grass, they find practical applications in their own world, such as the Greenlanders who use bam-boo poles to support their life-lines in blinding whiteouts and the companies in North America that manufacture lami-nated bamboo flooring.

In short, awareness of bamboo

Bamboo thrives in the warm, humid climate of the southeastern United States.

is spreading, and the material is increasingly used for decorative purposes in the West. *The Craft and Art of Bamboo* is the first Western book to explain techniques for designing and building with bamboo in language that interested professionals and amateurs can understand. To many of us, bamboo's cylindrical, hollow form is mysterious compared to the familiar squared, solid lumber used for most Western building purposes. This book is intended as a guide to help you create a wide range of home and garden items made with bamboo.

How did I begin my own love affair with bamboo? I first stumbled upon bamboo groves in the early 1980s while harvesting kudzu vines to weave into sculptural baskets. I was so awestruck by the giant, primitive stalks and dense growth, that for the first several encounters, I dared not even enter a grove. Then a friend took me to visit a 25-year-old grove along a slow-moving river in the southeastern United States. I was most impressed by the grove's boundless energy: its growth along acres of river bank; the vibrant, tall culms bursting with leaves that catch the sunlight; and the almost impenetrable ground cover of accumulated dead poles and branches leaning against each other in all directions. Not knowing exactly what to do with bamboo, however, I put it in the back of my mind and waited for an opportunity to work with it.

In early 1990, I was asked to create an environmental sculpture for an Earth Day celebration in Atlanta, Georgia. I envisioned a piece called the Earthball, measuring 10 feet (2.9 m) in diameter and constructed of natural, local materials woven together by Atlantans. The first challenge was figuring out what material to use for a spherical frame that would keep its shape while supporting hundreds of pounds of woven vines. My mind's eye returned to the long, green culms of bamboo. Experimentation proved what I expected: bamboo poles don't bend. Then I remembered a crude bamboo splitting device I'd seen workers using years before at the Bamboo Farm near Savannah, Georgia. We rigged up a similar device and split a truckload of culms nto splints. Tying the splints together into long, thin bundles, we successfully wove and tied the bundles into a spherical framework. The process and the result were magical, a celebration of working together with each other and the earth's bounty.

Energized by the Earthball, I was ready to further experiment with bamboo. I received grant money to explore bamboo as a new craft material for the West,

Above: Yotsume fence and gate by the Bamboo Fencer.
PHOTO BY DAVE FLANAGAN

Left: A snow-covered fence in Japan

Below left: Earth Day volunteers in Atlanta, Georgia hold the bamboo framework of the 10-foot-diameter (2.9 m) Earthball.

Below: A nanako border lines a walkway in Japan.

and soon after, in a moment of serendipitous timing, a Buddhist monk invited me to join him and a small group of Americans for three weeks' travel in Japan. Traveling through rural country-side and crowded cities, I saw bamboo—lots of it—made into everyday functional objects such as brush brooms, sup-ports for ancient tree limbs, and even long, bamboo tweezers for a shopkeeper to grab chunks of tofu! I saw bamboo fences, too. I was enchanted by the simple beauty of their construction and amazed by the variety of styles and applications used to build them. With the generous assis-tance of my Japanese guide and hosts, I returned to the United States with two all-pur-pose bamboo knives, ready to create with bamboo.

Months later, I held a reception to show the bamboo fences and screens I had made for an urban wildlife habitat. Guests came to celebrate bamboo, to drink *kakicha* tea from bamboo cups, eat *nori* rolls from bam-boo trays, and view an art form new to many of them. Since that time, I've continued to learn about bamboo. My vision is that as culms in the West continue to multiply, so will the number of Westerners growing and craft-ing with this most amazing plant.

Are you ready to get started? You can choose from projects in *The Craft and Art of Bamboo* to make contemporary structures for your landscape, garden, and home. You'll find that what appears strikingly beautiful is not always difficult! Most of the bamboo fences, screens, trellises, accessories, and other projects in this book are straightforward in construction once you've assembled the necessary tools and materials, and learned a few basics.

To further inspire and energize your creativity, each chapter contains images of the work of contemporary bamboo artists. They are the pioneers in this new field who felt an affinity with bamboo and applied themselves to learning about it by trial and error, observation, and research. Their work gives us a glimpse of the diversity that arises when a localized material becomes part of the global economy.

Relax in your favorite spot and learn about bamboo. Then visualize your landscape, garden, and home with your handcrafted works of art, integrating naturally into your nooks and crannies. Guaranteed to bring new energy to your environment!

Chapter 2

East Meets West, North Meets South

THE UNI-VERSAL ATTRAC-TION OF BAMBOO IS ITS BEAUTY. ALWAYS VIBRANT, ALWAYS ALIVE, ITS LEAVES RESPOND TO THE SLIGHTEST BREEZE. AFTER A RAIN WATER BEADS AND DROPS FROM THE LEAVES AND DOWN THE GREEN, GLASSY CULMS. ITS MAGICAL, SWISHING SOUND TRANSPORTS THE SPIR-IT TO A QUIETER PLACE.

Top left: Outside this entrance in Japan, bamboo-faced planters hold New Year's greenery of pine boughs and bamboo.
PHOTO BY KONOMO UTSUMI

Top right and far left: Bamboo scaffolding is being used during the construction of a Buddhist peace pagoda in Nepal.
PHOTO BY KONOMO UTSUMI

Left: Bamboo brooms for sale outside a shop in Tokyo, Japan.

Bottom left: Bamboo poles and rakes for sale at Prafrance.
PHOTO BY ADAM AND SUE TURTLE, ©*TEMPERATE BAMBOO QUARTERLY*, USED WITH SPECIAL PERMISSION.

Bottom right: Bamboo railings guide worshippers at a temple in Japan.

Beyond its beauty, the marvel of this magnificent plant is its abundance and adaptability. For centuries, the indigenous people of Asia, Africa, Australia, and North and South America have harvested and crafted its poles into shelter, tools, and other essentials of daily life.

Bamboo in the Eastern Hemisphere

Far Eastern culture is deeply interwoven with bamboo. It's so important to the material life of the region's cultures that it's inseparable from the fabric of those cultures. To the Chinese, bamboo is the proverbial symbol for resilience in adversity: tough yet pliant, bending without breaking, a symbol of humankind's moral imperative to maintain strength and fortitude in times of turmoil and hardship.

In Japan, traditional bamboo fences are seen on the grounds of temples, at cultural sites, and along the entryways of private homes. Many bamboo structures are less formal however, and are used for more functional purposes. In much of Southeast Asia, where bamboo is the equivalent of steel in the West, bamboo scaffolding is a common sight.

North America

In the warm, humid climate of the southeastern United States, grows the only bamboo native to North America, *Arundinaria gigantea*, or rivercane growing 30 feet (9 m) tall in dense stands called *canebrakes*. Native Americans used cane poles for roofs, barricades, arrows, and supports for squash and beans. They also wove split cane into mats and storage vessels. Today Cherokee Indians continue the tradition of harvesting, splitting, dyeing, and weaving bamboo into fine rivercane baskets.

South America

The Western Hemisphere's largest native species, *Guadua angustifolia*, or guadua, is among the world's strongest and most durable bamboos. Growing in great forests in the Andes Mountains, it has been a mainstay of Colombian culture for centuries. In regions where guadua abounds, dwellings are often built on earthquake-prone, mountainous terrain. A Latin American school of architecture that utilizes bamboo for urban and rural housing has emerged, and incorporates sturdy construction and sophisticated bamboo joinery techniques used by the general population for hundreds of years.

To the Chinese, bamboo is the proverbial symbol for resilience in adversity.

Above: In South America, guadua bamboo is frequently used as the framework for plaster, as shown in this house under repair in Colombia.
PHOTO BY FRANCISCO PLAZA

Far left: Traditional Spanish courting chairs allows an intimate conversation while a chaperone listens out of earshot.
DESIGN AND PHOTO BY YUCATAN BAMBOO

Left: Latin American architecture incorporates bamboo, an available and rapidly renewing resource.
PHOTO BY FRANCISCO PLAZA

Below: The bamboo of this coffee table and chair is heavily layered with an ebony lacquer. Design by Bruce Hanners.
PHOTO COURTESY OF YUCATAN BAMBOO.

Bamboo Farm and Coastal Gardens,
Savannah, Georgia

Located in the southeastern United States, this idyllic, 52-acre site contains the country's largest collection of bamboo open to the public. It exists as a publicly owned resource thanks in large part to the vision and generosity of two men: Dr. David Fairchild, and bamboo enthusiast and collector Barbour Lathrop.

The historic collection began in 1890 when three non-native timber bamboo plants were transplanted as ornamentals. The plants thrived in the sandy, rich soil and warm, humid climate, and after 15 years they covered an acre. The site was put up for sale, and botanist Dr. David Fairchild recognized its value. Fairchild contacted Barbour Lathrop of Chicago, a world traveller and collector of bamboo, who bought the bamboo plot and surrounding acreage. In 1919, Lathrop deeded the site to the U.S. Department of Agriculture as a Foreign Seed and Plant Introduction Station.

The original grove of huge timber bamboo that lines the road and entrance still anchors this unique coastal agricultural station. For more than 60 years, plant explorers from four continents have brought bamboo, ornamentals, fruits, and vegetables to be evaluated for their adaptability to the coastal southeastern United States. Thanks to gifts, purchases, and exchanges, the Barbour Lathrop Collection amassed and documented more than 150 varieties of bamboo.

The gardens were deeded to the University of Georgia in 1984, and the research and acquisition focus shifted away from bamboo. The groves, however, continue to thrive, their branches and leaves waving in the offshore breezes. Neatly contained in square and rectangular plots, the bamboo is divided by swaths wide enough for a tractor to pass through. The bamboo shade house, fences, and enclosures were built in the 1930s, and as poles decay, they are replaced as originally constructed.

Bamboo grows in square plots, divided by swaths wide enough for a tractor to mow.

In addition to housing, guadua is used in rural areas for telephone poles, animal corrals, and water pipes. On Colombian coffee plantations, guadua is crafted into large drying and storing structures and shade houses to protect growing crops from the sun.

East Meets West, North Meets South

Bamboo has been making its way into the West since the 1890s. Western plant collectors made expeditions to China and Japan and returned with exotic bamboo specimens to plant in the temperate zones of North America and Europe. Thanks to these and other introduced species, bamboo has spread, giving Westerners the opportunity to observe and learn about this amazing plant.

While bamboo inched its way onto American and European soil, the world's economies moved toward greater interchange and integration. Bamboo has emerged from its regionalized status, and its physical characteristics and ancient uses are being actively studied by Westerners searching for practical applications to Western needs. At the grassroots level, curious individuals are buying bamboo poles from local and Internet retailers, exploring how to adapt the

East meets west in this fence made of bamboo and milled lumber. Design by Regan Sheeley

material to their unique needs.

On a larger scale, scientists and engineers are searching for rapidly renewable building

Bamboo flooring, distinguished by its continuous grain and nodes, is available with natural, stained, finished, or unfinished treatments.

materials to substitute for wood. One result of more than a decade's experimentation is the manufacture of laminated bamboo board. The culms of timber bamboo, or another high-strength genus, are split put in a press, and crushed flat. After the green, outer skin and soft, inner walls are shaved off with a planer, the sheets are dried in a kiln, then glued and compressed in layers. A variation of the process is used to manufacture bamboo flooring. These high-tech bamboo laminates highlight the grain and visual interest of the material, and offer a durable, handsome alternative to wood.

Historic Bamboos at Biltmore
by Bill Alexander, Landscape Curator, Biltmore Estate

Upon entering the gates of Biltmore Estate near Asheville, North Carolina, visitors are immersed in the beauty of a landscape that covers more than 7,000 acres (2,800 ha) of hills and valleys stretching from the banks of the French Broad River. It's easy to understand why this setting was chosen for America's largest private residence, Biltmore House.

It's also hard not to be impressed by the extensive, shiny groves of bamboo that flank the entrance road and edge the towering forest. These groves came to exist and thrive in this noted landscape thanks to Frederick Law Olmsted, the man often referred to as the father of American landscape architecture. Olmsted's work at Biltmore Estate is one of the earliest documented examples of the extensive use of hardy bamboos in an American landscape. Today, more than 20 varieties of bamboo grow in more than 40 sites on the property.

Along with architect Richard Morris Hunt, Olmsted designed the estate in the 1890s for George Washington Vanderbilt, scion of one of America's most prominent and wealthiest families. Olmsted transformed thousands of acres of worn-out farms and parcels of slashed and burned woodlands into a park-like estate with miles of carriage drives, a productive working farm, and extensive gardens and pleasure grounds.

A master of naturalistic design, Olmsted respected the special qualities of the site and the character of the nearby Blue Ridge Mountains. He employed local, indigenous plants as the backbone of his palette, but also used several non-indigenous plants to enrich the natural-looking scenery. Several species of bamboo were among his choices. Drawing from the English landscape gardening style known as the *picturesque*, Olmsted aimed to heighten nature's mysterious and bounteous aspects. He did so by incorporating

bamboos with other plantings, adding tints and textures of foliage and a constantly changing play of light and shadow. Profusely planted masses of bamboo provided richer, lusher growth than nature might produce unaided.

Olmsted also took advantage of bamboo's numerous forms and growth habits, creating a layered effect. Bamboos were utilized in mass with other evergreen plantings in the understory and foreground to suggest a feeling of "sub-tropical luxuriance" in a landscape that, in Olmstead's words, would make northern guests "feel that they are decidedly nearer the sun." Some of his choices included the diminutive, fine-leafed pygmy bamboo (*Sasa pygmaea*), that grows only one or two feet (30.5 to 61 cm) tall; the shining, palm-leafed bamboo (*Sasa palmata*) which reaches several feet in height; and the towering forms of the "timber" or "grove" bamboos of the genus *Phyllostachys*, which grows more than 40 feet (12 m) tall.

Yucatan Bamboo

At the turn of the twenty-first century, deep in the jungles of Mexico's Yucatan Peninsula, the first mature culms of *Dendrocalamus strictus*, iron bamboo, were harvested from a converted nineteenth-century sisal plantation. This variety of giant, clumping bamboo had never before grown in the Western Hemisphere. Iron bamboo is well-named, possessing an almost solid culm that makes its poles among the strongest in the world.

In 1995, the founder of Yucatan Bamboo, Bob Gow, had a rare opportunity to secure iron bamboo seeds from a source in India. The bamboo had just flowered and would not do so again until sometime around 2045. Gow planted 22,000 seedlings, and with his first harvest, joined the ranks of environmentally conscious businesses that are growing, harvesting, and selling bamboo as a rapidly renewable resource. At Yucatan Bamboo, the 3- to 5-inch (7.6 to 12.7 cm) diameter culms are milled to form rectangular or square lengths that can be nailed or screwed without cracking or splitting.

The thriving grove is now the largest commercial bamboo plantation in the Western Hemisphere. Covering hundreds of acres, the bamboo grows in rows along stone roadbeds that date from the pre-Columbian Mayan empire. More than 70 Mayan Indian villagers work at the plantation and craft the bamboo into furniture, fences, and other products sold in western markets.

Above: Harvested iron bamboo poles
PHOTO BY ALEX BOND

Top left: Restored Hacienda Xixim sits in the midst of the largest commercial bamboo plantation in the Western Hemisphere.
PHOTO BY ALEX BOND

Second photo from top left: A Mayan craftsman displays his work.
PHOTO BY ALEX BOND

Bottom left: Mayan workers wash fence panels before export to Western markets.
PHOTO BY KAREN WITYNSKI

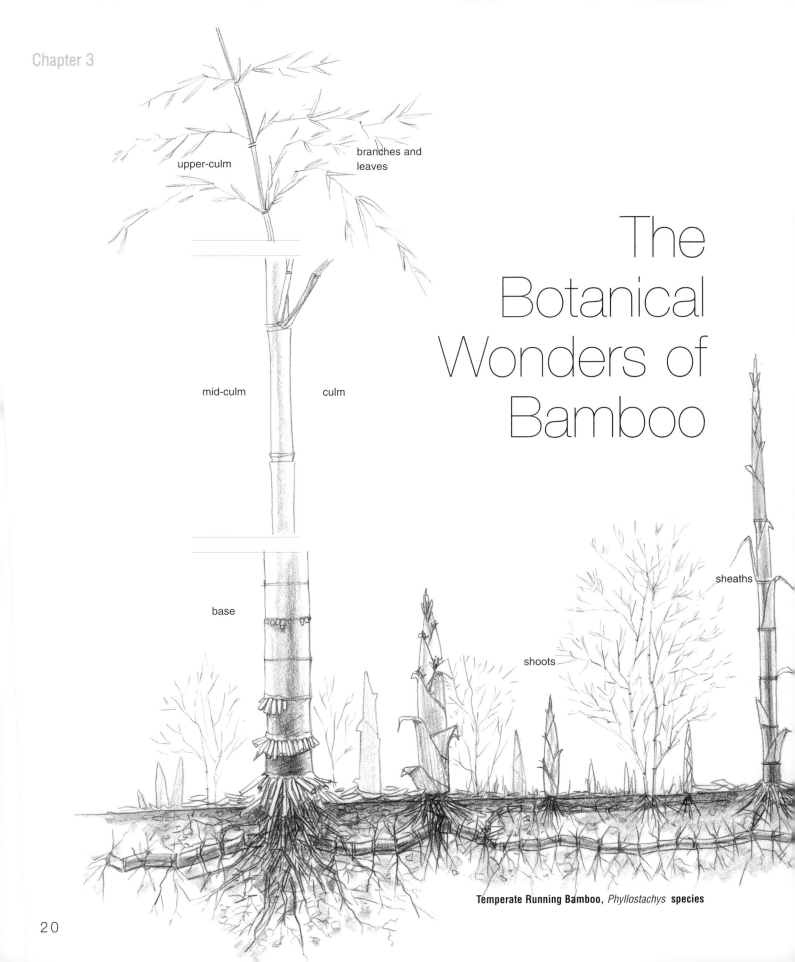

upper-culm

branches and
leaves

mid-culm

culm

base

shoots

sheaths

The Botanical Wonders of Bamboo

Temperate Running Bamboo, *Phyllostachys* **species**

shoots

rhizomes and roots

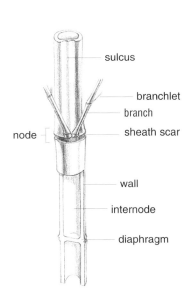

sulcus

branchlet

branch

node

sheath scar

wall

internode

diaphragm

BAMBOO IS A LARGE SUB-FAMILY OF THE GRASS FAMILY *GRAMINEAE*. A GIANT PERENNIAL GRASS, BAMBOO HAS A HARD, WOODY, JOINTED STEM. ITS FAMILY, *BAMBUSACEAE*, IS DISTINGUISHED FROM OTHER GRASSES BY ITS FLOWERING CHARACTERISTICS AND ITS SPECTACULAR GROWTH RATE: BAMBOO REACHES ITS FULL HEIGHT AND DIAMETER IN ONLY ONE OR TWO SEASONS!

16"
(40.6 cm)

15°

Bamboo flowers unpredictably, often inconspicuously, once every 30 to 120 years.

The 1,200-plus bamboo species are classified according to the way their underground structures, called *rhizomes*, grow and spread. The rhizomes of *clumping bamboo* are bunched together, and buds shoot up when rainfall comes after a dry season. Clumping bamboo grows in tropical, equatorial zones in dense circular stands, multiplying outward. The rhizomes of *running bamboo* send up buds in all directions in the spring, when the sun's warmth returns to cooler, temperate climates.

How Temperate Bamboo Grows

The growth spurt of the stem, or *culm*, is bamboo's most spectacular characteristic. Temperate bamboos reach most of their height within one month and are full grown by the second. Amazing, considering the largest temperate bamboos reach 60 feet (18 m) in height!

This dynamic growth is anchored and nourished by an underground network of *rhizomes*, the food storage structure from which roots, buds, sprouts, and more rhizomes grow. Segmented like the culm,

rhizomes form a mat so dense that, even today, people find refuge from earthquakes in centuries-old bamboo groves, knowing the rhizomes will hold the earth together. The rhizome network is shallow, with 80 percent lying in the top 12 inches (30.5 cm) of soil and the remainder in the next 12 inches.

Fibrous *roots* grow from the nodes of the rhizome, seeking out and delivering water to rhizomes and culms. Pointed, tough *buds* grow from rhizomes, pushing their way to the surface to emerge as *shoots*. A husklike, tubular grass sheath surrounds each shoot. The *sheath* protects the growing shoot and culm while it's still young and soft, and its overlapping leaves wrap the culm so tightly but with minimum friction that it forces the culm to grow upward. Shoots grow telescopically and with great speed to become culms. When they emerge from the ground, the base of the shoot already measures the same diameter it will have as a mature culm.

Bamboo culms grow stiff and upright. As the walls become woody enough to be self-sup-

porting, the sheaths drop away. Culms are widest at the base, tapering upward to a thin tip. They may arch over or grow upright, depending upon their place in the grove. Depending on their species, some culms may bend abruptly or at zigzagging angles at their base. Culm *walls* are made of tight cellulose fibers containing *lignin* (also found in wood). The outer walls' strength and hardness is due to a silica content that can reach 5 percent.

As shown in the illustration on page 21, *nodes* are the raised joints that segment the culm into hollow compartments. Buds and branches emerge from nodes primarily in the upper culm. Immediately below the nodes are *sheath scars*, which mark the previous attachment of sheaths. Inside the culm is the *diaphragm*, a rigid membrane that connects the walls and gives strength to the hollow cylinder.

Internodes are the hollow segments of the culm between nodes. Close together at the base, internodes elongate gradually to their maximum length in mid-culm. Then, as branching begins, the spaces decrease toward the tip. The *sulcus*, a lengthwise groove in the upper internodes, is formed by an extrusion process as the culm pushes past a branch bud held tightly in place by the sheath. The inside of the internode walls are covered by a cream-colored, papery material.

Segmented *branches* emerge just above the nodes on alternate sides along the culm, and *branchlets* grow from the branches. These stalklike projections hold leaves and leaf sheaths, giving them flexibility to withstand the elements. From the branchlets grow bushy, abundant leaves. The upper surfaces of the narrow, elongated leaves are a smooth bright green, the undersides a powdery, matte green. Bamboo plants are evergreen, gradually dropping leaves through fall and winter, with new leaves appearing in spring.

As the culm ages, its cellular tissue hardens. Within three to five years, the culm is ready for harvest, though its total life span ranges from five to eight years. Within this time, a grove becomes established. In an optimal climate and site, if harvested and groomed appropriately, the grove will mature to produce culms of maximum size.

Bamboo flowers unpredictably, often inconspicuously, once every 30 to 120 years. Flowering saps the energy of the grove but doesn't necessarily mean the death of the plant. The culm may die back, but the plant will most likely grow back

from its rhizomes or seeds. Roughly the size and appearance of a grain of wheat, bamboo seeds germinate under the leaf litter, and seedlings may eventually re-carpet the grove.

Containment and Maintenance

Although considered invasive, running bamboo is better termed "vigorous" and can, in fact, be contained by natural barriers such as waterways and well-traveled paths and roads. Poured concrete and sheets of alu-

minum or heavy molded plastic make contained plantings. Barriers should be set 24 to 30 inches (61 to 76.2 cm) deep around a planting, extend 2 inches (5.1 cm) above the soil, and angle outward. Sheets must be overlapped and tightly joined with screws; caulk if there is any space for rhizomes to penetrate. All bamboo, contained or not, needs regular maintenance. Remove unwanted shoots in the spring, harvest mature culms annually, and remove dead and decaying culms.

A grove of tea-stick bamboo is bordered by water, a natural barrier to undesirable growth.

Chapter 4

Harvesting, Purchasing, and Preparing Bamboo

Very few of the groves in the southeastern United States are properly maintained; the majority run rampant.

THE BEAUTY OF GATHERING YOUR OWN BAMBOO IS THE SATISFACTION GAINED FROM HANDLING A NATURAL MATERIAL AND CRAFTING IT FROM ITS RAW TO FINISHED STATE. OF COURSE, HARVESTING GIVES YOU A GOOD REASON TO EXPLORE A BAMBOO GROVE, WHICH IS A WONDERFUL ADVENTURE IN ITSELF. TIME REVERSES IN THE DARKENED INTERIOR OF THE GIANT GRASS FOREST, WHERE QUIET, SOLITUDE, AND THE GENTLE FLUTTERING OF LEAVES SURROUND THE HUMAN VISITOR.

Harvesting

If you have access to a bamboo grove from which you can harvest your material, consider yourself lucky, but be prepared. Harvesting bamboo is for those who enjoy getting physical and interacting with nature in its unbridled state, and those who are willing to encounter anything in a grove from poison ivy to old refrigerators.

Getting Prepared

Before embarking on your expedition, be sure to obtain permission from the property owner. Next, assemble your harvesting tools (see the adjacent photograph), and wear long pants, a long-sleeved shirt, gloves, sturdy shoes, and eye protection. Don't underestimate the value of covering your arms. One hot spring day, I wore a T-shirt to harvest and haul bamboo, then spent the night nursing a fiery red rash on my forearms. Tiny silica crystals on the surface of new growth culms had pierced my skin like fiberglass. Ouch!

When to Harvest

Winter and very early spring are the natural times of year to harvest temperate bamboo. Growth is complete for the year, and roots and culms are in a resting cycle. Your work will be easier, too. Vines and overgrowth amidst the culms have

Harvesting tools include a collapsible tree pruning saw, machete or loppers, and hand pruners. Bring rope, bungee cords, collapsible sawhorses, and a tape measure too.

lost their leaves, snakes and insects are dormant, and the air is cool and brisk.

Selecting Older Culms

The underlying principle of harvesting bamboo from a grove is the same as maintaining a healthy garden: prune out old growth to make way for the new. Generally, the culms of temperate-climate, running bamboo grow old and die after five to eight years. It's best to harvest older, mature culms because their cellulose walls have developed strength. Also,

Older culms may be yellowed or scarred.

Another technique is to look upward in the grove canopy for dying, leafless tops. They're not always easy to see, however, and it can be difficult to distinguish which top is connected to which culm. In this situation, it's very helpful to work with a friend. While you're inside the grove, have your spotter positioned outside to direct you to areas with dying tops. Then shake the culms until your friend confirms that you're shaking the one with the leafless top. You can also harvest standing, dead culms. Just check to make sure they're not cracked, split, or rotten.

Cutting and Hauling

Once you've located an appropriate culm to harvest, check to make sure that you have a viable "escape route," a straight path out of the grove starting from the base of the culm. Take into account that both you and the 12- to 40-foot-long (3.6 to 12 m) culm may have to squeeze and maneuver through a thick bamboo forest.

much of the sugar in the walls has converted to starch, making them less attractive to boring insects. Older culms also contain less moisture, making them lighter to transport.

How do you distinguish old culms from new in a grove? Look closely. If a culm has a bright green, relatively clean surface and dried sheaths encircling its base as shown in the above photograph, it's probably a first- or second-year culm. If the culm is dark or dull green, perhaps with brown, yellow, or tan areas, chances are it's older and ready for harvest. Quantities of mildew and dirt on the culm are also a clue to older age.

After selecting the culm and confirming that you have a way out, crouch down and clear away the *duff*, the dried and decaying bits of bamboo leaf, sheath, and stalk, from the base of the culm. Use the coarse-tooth pruning saw to cut the culm as close and even

with the ground as possible. This practice eliminates stubs that could trip you, stick you, or hold water that breeds mosquitoes. If the base is 3 inches (7.6 cm) or greater in diameter, first make a back cut about one-third of the way through to get a clean cut and to keep the saw from getting stuck.

After cutting, fold up the saw and secure it in your pocket. Haul the culm, base first, through the escape route. Haul the culm into the open where you've positioned two sawhorses. Remove branches with a machete or loppers. If you've already planned your bamboo project and its components, you can measure and cut the lengths you need. Otherwise, a good rule of thumb is to cut the longest length your vehicle and storage area will accommodate.

Grove Etiquette

More often than not, the branching top third of the culms will not be used in your projects. What to do with all that brush? Since branches and culms break down slowly, leaving brush to decay haphazardly in the grove is an unhealthy environmental practice. The debris restricts the air movement necessary to ward off fungus and insect attacks. It also hampers other visitors' efforts to harvest and tend the grove. For long-term bamboo brush com-

posting, it's best to designate an area adjacent to the grove for the orderly stacking of brush and culm debris.

Transporting

Moving bamboo to your work space can be tricky. Unlike milled lumber, bamboo slides and rolls, making it a hazard on the highway if it's not snugly tied to your car. Once I was in a hurry to drive a short distance from grove to workshop, and I did a sloppy job of lashing 12-

foot (3.6 m) lengths of bamboo onto my car's roof rack. As I drove downhill, the poles slid off, and I watched with horror as they went flying like torpedoes onto the street ahead of me. Lesson learned. Equally surprising is the weight of green bamboo, so be prepared for that too. On one expedition, the roof of my tiny station wagon caved in under the weight of my harvest, and I agonized every time my car roof thumped in and out for the 75 miles (120 km) home.

Bundling is an effective remedy for these challenges. As shown in the photographs to the left use a rope or several bungee cords to tightly wrap together a manageable number of poles. Create bundles that you can lift and carry by yourself or with a friend's help. Tightly secure the bundles to your vehicle and to each other. Use lots of rope and double-check that the load won't shift.

Purchasing

If you don't have access to a grove, you can buy bamboo poles from a number of sources. For small projects, you can often find 4- to 5-foot (1.2 to 1.5 m) bamboo stakes, ½ inch (1.3 cm) in diameter, sold singly at import stores, garden centers, and home improvement stores. For larger projects,

there are many sources for poles. See chapter 10 for suppliers.

Western companies import a variety of sizes of bamboo poles, primarily *tonkin (see page 33)*, for the nursery and decorative trades. Imported from China and southeast Asia, the poles have been sorted, sized, bundled, and fumigated before they reach Western warehouses. Poles 1 inch (2.5 cm) or less in diameter are sold in bundles of 50 or more, while poles 1 to 3 inches (2.5 to 7.6 cm) in diameter are sold in bundles of 5 to 25. The largest poles, 4 to 6 inches (10.2 to 15.2 cm) in diameter, are often sold singly. Many suppliers offer splitting and cutting services.

As non-native species of bamboo have established themselves in Western groves, a variety of mature culms are now being harvested annually, primarily running bamboo in the United States and Europe, and clumping bamboo in Australia. Poles grown, harvested, and shipped within domestic borders are not required to be fumigated. Some growers cut to order and ship green.

Although not made of bamboo, *reed fencing* is often carried by bamboo importers. Stalks of heavy grass are bound together with wire and are sold in 8-foot (2.4 m) rolls in widths of 4 and 6 feet (1.2 to 1.8 m).

If you buy by mail or on the Internet, be sure to factor in shipping costs when placing your order. In the United States, parcel delivery services will deliver bundles up to 8 feet (2.4 m) long. For shipments exceeding these dimensions, shipping is by common carrier, i.e., trucking companies.

Storing

Bamboo should be stored out of the weather in a covered area such as a garage, shed, or dry basement. Use pallets or sawhorses to keep the poles off the ground to prevent moisture from wicking up and causing rot. I have used a variety of found materials to create a wall of bins, baskets, and containers to organize and store my bamboo. Keep in mind that the greater the variety of bamboo you accumulate the more choices you have during the creative process. Collect, collect, collect! Save your scraps. Keep your bamboo stored where you can admire it. Let it inspire you.

Drying and Weathering

As freshly harvested bamboo dries, the chlorophyll fades and the uniform forest green may become shades of green and

yellow with streaks of brown. Scarring, caused by culms falling across one another, or raised dark spots may be noticeable. These harmless sprinkles, known as "sesame," are caused by organisms that alter the surface of the outer wall.

After about six months of outdoor exposure, bamboo dries completely, turning a uniform light tan. As bamboo structures are exposed to the elements, they naturally become weathered. Sun and rain break down

the outer layer of silicone, and the surface becomes pitted. Mildew settles in and dirt accumulates.

Annual cleaning and sealing greatly increases the life and appearance of bamboo structures; see chapter 5 for suggested annual maintenance. If, however, you choose to allow an outdoor bamboo structure to age naturally, it will no doubt become one with its surroundings, returning gracefully to the earth.

You can attach lengths of lumber to walls or ceilings to store poles efficiently.

As fresh bamboo cures, its color changes from green, to brown, to gray.

Insect Damage and Control

Where there is food, there are bugs! The sugar in the walls of freshly cut bamboo may become a meal ticket for wood-boring insects. The mature insects deposit their eggs in the harvested culm. As the eggs hatch, the larvae chew through the inner walls and finally emerge through the outer wall.

Fumigation, best done in industrial settings, is the only sure way to prevent and eliminate bugs. The risk of infestation can be lessened however, if you harvest in late winter, harvest mature or dying culms, and store harvested poles upright with their base ends down until drying is complete. All three practices help keep the sugar in the poles at their lowest level while curing. Sweating the poles will also reduce the risk of insect damage; see chapter 5 for this technique.

If your stored bamboo becomes infested, discard the worst. Poles or other pieces that are slightly infested can be used for supports, markers, and simple temporary structures lasting only one or two seasons. If borers are present in poles or projects you've already completed, you can kill the larvae with chemicals or by freezing them. Put short lengths in plastic bags and place them in the freezer overnight. This kills the chewing larvae but not the eggs, so you'll need to repeat the process several weeks later after any eggs have hatched. For long poles or large, finished projects, chemical insecticides for boring insects are effective and widely available in garden

Over time, the sun may dry and bleach bamboo to a silver gray. In humid climates, moss and lichen may take hold.

Pinprick-size holes in the bamboo and tiny mounds of finely ground powder are clues to insect infestation.

stores. Follow the package directions carefully to minimize your exposure to the chemical and its impact on the environment. Work outdoors, and wear protective gloves and a respirator. Brush the chemical solution onto the outside walls of the infected lengths. If the bamboo is split, apply it to the pulpy inside walls. Let dry overnight, then apply sealer.

Cleaning and Fungus Removal

Whether bamboo is harvested or purchased, washing with soap and water removes dirt and mildew from bamboo and brings out the luster of its shiny surface. To remove the worst of the grime, line up poles against a wall or across two sawhorses, and use a hose with a high-pressure nozzle to blast off the dirt. Spray with a nonsudsing, mildew-removing cleaner, then follow with a water rinse and let dry.

When bamboo is surrounded by stagnant air and dampness, patches of mold and fungus may grow on its surface and on the inside walls of split bamboo. To discourage such growth, regularly rotate your materials. In closed spaces, use a dehumidifier to collect moisture and a fan to keep the air

For a more thorough cleaning, especially around nodes where dirt collects, use a scrub brush with the cleaner.

moving. You can spot-clean mold and fungus by moistening a cloth with isopropyl alcohol and rubbing it over the affected area. Alcohol kills the organisms, disinfects the area, and evaporates quickly.

Bamboo Poles Available for Purchase

COMMON NAME/ GENUS & SPECIES	Golden *Phyllostachys aurea*	Vivax *Phyllostachys Vivax*	Red Margin *Phyllostachys rubromarginata*	Black *Phyllostachys nigra*	Robert Young *Phyllostachys viridis 'Robert Young'*
TYPE	temperate, running	temperate, running	temperate, running	temperate, running	temperate, running
CULM DESCRIPTION	culm is green, turns golden yellow in sun; often distinctive, ornate compressed nodes at base; some culms may zigzag at the base	unpronounced nodes; resembles Giant Bamboo but internodes are slightly off-center and somewhat curved; old culms mature to light grey-green	smooth, medium to dark green, long internodes of 16 inches (40.6 cm) or more; in larger culms, walls bulge asymmetrically above node	slender, chocolate brown to ebony black; green the first year, then turns dark	initially sulphurous green, turns gold-yellow by end of year; random green stripes along internodes
WOOD QUALITY	not of superior quality, but good for general purposes	very good quality	very good quality	thin-walled but durable; used to make furniture and decorative objects	very good
ORIGIN	China	China	China	China; also introduced to Japan long ago	China
DIAMETERS AVAILABLE	½ to 2½ inches (1.3 to 6.4 cm)	2 to 3 inches (5.1 to 7.6 cm)	1 to 2¼ inches (2.5 to 5.7 cm)	½ to 2½ inches (1.3 to 6.4 cm)	2 to 3 inches (5.1 to 7.6 cm)

Iron **Dendrocalamus strictus**	*Tropical Black* **Gigantochola atroviolacea**	*Tonkin* **Pseudosasa amabilis**	*Henon* **Phyllostachys nigra, 'henon'**	*Moso* **Phyllostachys pubescens**	*Giant Timber* **Phyllostachys bambusoides**
tropical, clumping	tropical, clumping	tropical, clumping	temperate, running	temperate, running	temperate, running
slender, thick to almost-solid wall	black with occasional faint green strips that turn tan when dry	straight with smooth nodes, cures to blond or tan	mature culm covered with green/grey waxy coating; may be removed to reveal a rich hue	can grow large, up to 7 inches diameter	straight, even, thick-walled
hard, dense, important for construction	sturdy, used for fine furniture and musical instruments	thin-walled but tough, exceptionally resilent and strong	thin-walled but strong; almost as good as giant timber bamboo	wood relatively soft, but poles are very versatile	hard, versatile, used in construction and crafts
India	Java, Sumatra	China	China; also introduced in Japan long ago	China; also introduced to Japan long ago	China, India. Also introduced to Japan long ago
1 to 1½ inches (2.5 to 3.8 cm)	2 to 4 inches (5.1 to 10.2 cm)	¼ to 2 inches (6.35 mm to 5 cm)	2 to 3 inches (5.1 to 7.6 cm)	2 to 7 inches (5.1 to 17.8 cm)	1 to 5 inches (2.5 to 12.7 cm)

Chapter 5

Tools, Materials, and Techniques for Working with Bamboo

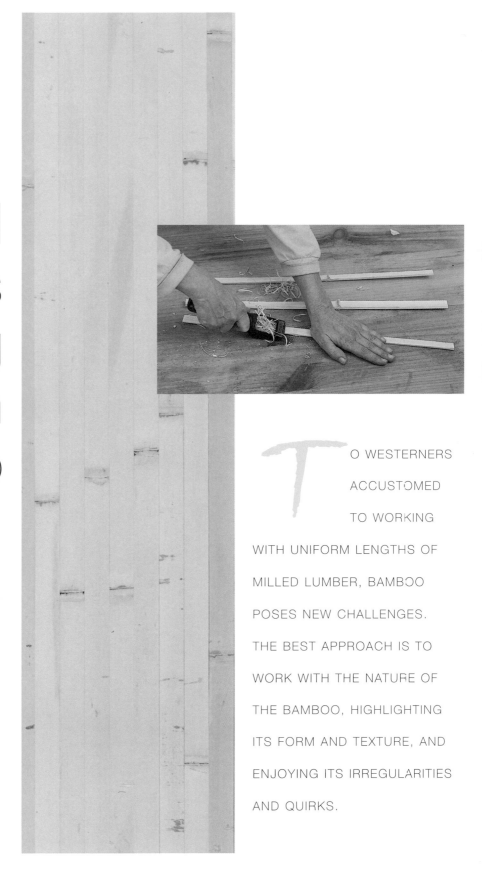

T O WESTERNERS ACCUSTOMED TO WORKING WITH UNIFORM LENGTHS OF MILLED LUMBER, BAMBOO POSES NEW CHALLENGES. THE BEST APPROACH IS TO WORK WITH THE NATURE OF THE BAMBOO, HIGHLIGHTING ITS FORM AND TEXTURE, AND ENJOYING ITS IRREGULARITIES AND QUIRKS.

Planning and Design

Start by choosing a project that interests you, suits your landscape, garden, or patio, and fits your available time. Building with bamboo can be surprisingly labor intensive, and large projects should be well planned. For any project, calculate the number of poles needed and account for their diameter and length just as you would with lumber. Figure in extra poles to replace the unsuitable ones which may be crooked, split, or otherwise unusable. When making trellises or other structures, lay out your design on a large flat surface such a patio, and adjust and replace poles to get the best design and fit before you cut.

Cracking

Know that bamboo culms can sometimes split when their walls dry and the air trapped inside expands with changes in temperature and humidity. I cringe every time I hear a "BOO!" coming out of my studio. Accept it: cracking happens. There is a technique, however, that may lessen the chance of splitting. Before assembling a project, use rebar or threaded pipe to knock out node membranes inside the culm. This gives heated air inside the chambers a place to escape.

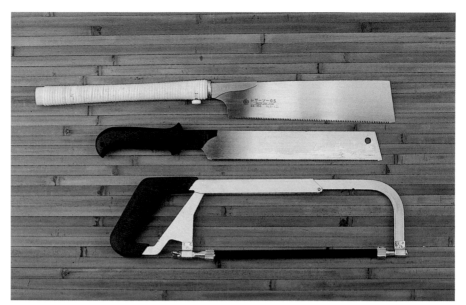

Top to bottom: Japanese saw, plastic pipe saw, hacksaw.

Cutting Bamboo

Bamboo has a smooth, hard outer surface. Tiny silica crystals form a glasslike coating which repels rain but is difficult to penetrate. Beneath this outer wall is an inner wall of tough cellulose fibers. Your equipment and cutting technique will need to account for these factors.

SAWS AND SAWING TECHNIQUES

To cut directly across a culm, you'll need a sharp saw. Many types of saws will work, but for rough cuts I rely primarily on a hand saw designed for pruning trees. Fine cuts are best made using a pull saw with a thin blade and 22 teeth per inch (2.5 cm). The cutting teeth must saw into the fibers cleanly, so as to not shred the culm's sur-

face. It's essential that the saw teeth be sharp and complete, no easy task since bamboo's hard silicone surface quickly dulls blades and damages teeth. Western hacksaws with metal-cutting blades work well, as do Japanese saws, which are more heavily weighted with blades specifically designed to cut bamboo. Power saws also work well, and the fine teeth of a ceramic disc blade give a good cut.

It's much easier to saw cylindrical bamboo when you use quick-release clamps. Use the clamps to secure bamboo to your workbench, and making precise cuts will become a no-struggle, no-juggle situation. Another approach is to use a Japanese cutting box. Designed for use while sitting

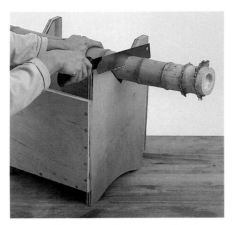

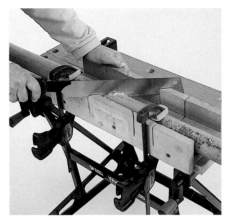

Photo 1: When using a Japanese cutting box, rotate the pole to get a clean cut.

Photo 2: To eliminate "feathering," hold the end of the bamboo while you cut.

Photo 3: Using a miter box to cut bamboo at a 45- or 90-degree angle.

on the floor, the box can also be placed on a bench to achieve a comfortable working height. The v-shape of the box provides a secure groove for the culm as shown in photo 1. When not in use, the box holds tools.

If your saw is too coarse or dull, the outer wall of bamboo you cut may shred or "feather." As the saw finishes the cut, the culm may break too soon, leaving a ragged edge on one side of the cut and the inner wall exposed on the other. For initial

Western-style "push" saws such as hacksaws are primarily "push to cut," meaning the teeth cut on the stroke moving away from your body.

Japanese bamboo saws are "pull to cut," cutting on the stroke moving toward your body.

harvesting or rough cuts, feathering may not matter. For fine cuts, however, you'll want to achieve a clean, sharp edge. If you're using clamps, saw about three-quarters through the culm. Next, use your other hand or a support to carefully hold the extended end (see photo 2). Continue to simultaneously saw, while supporting the extended end, until the blade cleanly penetrates through the outer layer.

If you're using a cutting box, first cut through the hard outer layer and the wall. Then, as you saw, rotate the bamboo so the saw is always cutting into the fibers. This helps you prevent tearing the outer fibers. Keep the blade straight so that when you complete the rotation the beginning and end of the cut will meet.

You may wish to cut 45° angled ends, which give an open, elegant feeling to a project. Also, if

you wish to butt two pieces of bamboo against each other at a 90° angle, it's important that your cuts be correct. To avoid problems, use a simple miter box, which is an open-ended "jig" marked with slots set at 45° and 90° angles to guide your saw. Mark the desired angle on the bamboo, and place it in the miter box. As shown in photo 3, line up the mark with the appropriate set of slots, clamp or hold the bamboo securely, and saw. For projects requiring less uniformity or precision, diagonal cuts can be made without a miter box, but it always helps to mark the diagonal angle before you cut.

USING A JAPANESE KNIFE

A Japanese knife is an excellent tool to "manicure" bamboo when you need to make fine adjustments in circular or square openings in order to fit poles into those joints.

Japanese knives and the technique for using them is quite different from the Western-style pocket knife. The blade is just 1 inch (2.5 cm) or so long and is placed on the diagonal at the end of a steel handle.

To cut with the knife, hold it near the blade in your dominant hand. Grip the bamboo with your other hand, near the hole you wish to scrape. If the pole is long enough, tuck it under your arm against your side to stabilize it. Now use the thumb of your non-dominant hand to push the knife

Japanese knives for splitting and cutting.
KNIFE SHEATH ON RIGHT BY TAKEI AKINORI

Photo 4

down into the fibers. Unlike Western knife techniques, the hand holding the knife applies no forward pressure. Instead, the thumb pushes downward, lending more control of the very sharp blade (see photo 4). When you've cut around one-quarter of the curve, you'll notice that the blade will not properly cut through the next quarter curve. To continue carving, you'll need to readjust the position of the pole, your hands, or the knife so the blade cuts through the fibers.

Splitting Bamboo

Artistic possibilities expand when you work with split bamboo. The halved side can be placed against flat surfaces, and shorter split lengths make

good wall coverings and inlay for bamboo "mosaics."

Freshly harvested green bamboo is easier to split than dried and cured bamboo. Green bamboo shrinks, however, and

Split bamboo is woven between bamboo stringers. Design by The Bamboo Fencer.
PHOTO BY DAVE FLANAGAN

37

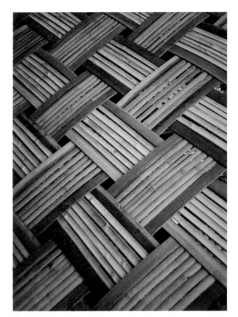

Detail of a split and woven bamboo fence panel.
DESIGN AND PHOTO BY THE BAMBOO-SMITHS

The front of this shed is faced with split bamboo and tree bark.
DESIGN BY MICHEL SPAAN

split green bamboo shrinks even faster. It's easiest to make pegs, handles, and other fittings from green bamboo, but allow them to dry thoroughly before incorporating them into a project.

SPLITTING TOOLS AND TECHNIQUES

It's easy to split bamboo by using a splitting knife and a rubber mallet. A good splitting knife has a strong blade and a sharp double bevel. Splitting knives vary in size and heft; use a blade 8 inches (20.3 cm) long to split large-diameter and woody lengths, and a shorter, thinner blade for finer work. An all-purpose splitting knife is the Japanese *take ware*, with a blade 8 inches (20.3 cm) long, hand-forged of high-carbon

steel. Any strong, double-beveled knife works well as long as it's kept sharp. You'll also need a rubber mallet to drive the knife through the culm. Don't be tempted to substitute a steel-headed hammer; it will damage the top of the splitting blade.

Successful splitting requires two things: good initial placement of the blade, and keeping the blade along the same vertical grain throughout the length of the culm. First, cut the pole to the length desired. Next, determine which end you will start your split. The accepted school of thought is to split from the top of the culm toward the base, "top to bottom always." On mid-culm lengths, however, it can be difficult to

determine which end is which; in that case, look and feel for the sheath scar of the node. It's always toward the base.

To determine the placement of the split, put one end on the floor or a low surface and eyeball the length from top to bottom. Rotate the pole; it may be very slightly curved. The idea is to split along the axis of the pole, which will yield two somewhat straight lengths, as

A worker at the Bamboo Farm, Savannah, Georgia splits a pole through a blade welded onto an angle iron.

opposed to two curved lengths. The split halves may arch slightly but will flatten when attached to a hard surface with lashing, screws, or nails. At the top end of the pole, mark the point where you plan to begin the split.

To split, stand the pole on end on a chopping block or brace it in a corner. Place the knife blade over the starting mark you made, with the end of the blade protruding several inches beyond the pole. Using the rubber mallet, knock the blade securely into the culm (see photos on right). The force of the strike and the thickness of the walls will determine the depth of the split. If the pole is thick-walled, the blade may drive into the grain only ¼ inch (6 mm) or so. If the culm is thin-walled, the blade may go deeper.

Using the mallet, hammer the blade through the full length of the culm, keeping your eye on the split to keep the two sides equal. If they start to split unevenly, adjust the angle of the blade by gripping the handle and twisting the sharp edge of the blade toward the side that needs to be thinner and away from the side that needs to be thicker. Maintain this angle as you hammer the blade through several nodes or until the sides become equal again. Then straighten the blade and continue splitting. Don't expect

Clockwise from top: machete, splitting knife, *take ware*, rubber mallet, four-way splitter.

perfection every time! Some poles are tough or seem to have minds of their own. Practice is always the best teacher.

Some artisans skip the mallet and use only a knife and a pounding or "tapping" action. Position the blade across the end, and with one sure and swift motion of the wrist, lift the

blade and drive it securely into the culm. With the blade now fixed in the bamboo, tap the culm's base end on a hard surface. At each impact, the blade will cut through the culm. You can vary the stroke from a light tap to a hard pound. Angle the blade to achieve even halves.

Making Splints and Removing Pith

Splints are narrow lengths of bamboo used to make pegs, chopsticks, woven fence panels, mats, and baskets. To make splints of any width, use a splitting knife and mallet to split a length of bamboo in half. Remove the diaphragms with a hammer or knife. Split the half-round length, and split each half again. Repeat until the splints are the desired width.

You can use a hammer to knock out diaphragms.

To make splints of ¾ inch (1.9 cm) wide or less, you'll need to have more control over the splint. The Japanese technique of pinching the splint while twisting the knife can help. Near the top of the splint, hold the edges between your thumb and index fingers. If the splint is long enough, tuck the other end under your arm against your chest to steady it. Hold the splitting knife in your other hand and rock it back and forth on the end where you want the split, pushing the blade in so it

When splitting or removing the pith to make splints, wear leather work gloves until you've mastered the techniques. A blade that's sharp enough to cut bamboo can easily cut your fingers.

separates the fibers. Lift it out, squeeze the splint between your thumb and index finger, and with a short, quick flick of the wrist, embed the blade further into the fibers. Release the "pinch" and twist the knife blade. This separates the fibers and allows you to move the knife further down the splint. Continue to pinch, strike, and embed the blade, and twist, keeping the split along the same grain.

To achieve a finer, more flexible splint, you can remove a third or more of the pithy inner wall. Holding one end of the splint,

You can dry and store fine strips for later use. Soaked in warm water for 10 minutes, they can be used as lashings or weavers.

place a sharp splitting blade lengthwise across the other, top end. Rock it back and forth to embed the blade into the pulp, then twist it, breaking apart the sheet of pulp along the grain. Continue to top the blade down the splint and twist. If the splint is still not fine or flexible enough, remove another sheet of pulp. Use a knife or shaver to smooth the inner wall and edges.

You can also use a commercial splitter; it's made of four or

Splitters like the one shown are used extensively in commercial bamboo processing throughout China and Southeast Asia.

more razor-sharp, hardened steel blades affixed in a cone shape that's attached to two handles. The choice of the number of blades depends upon the diameter of the bamboo and the desired width of the splints. To use the splitter, center it at the thinnest end of the culm. Tap it or use a rubber mallet to drive the blades into the culm. Firmly grip the handles and tap the bottom end of the culm against a hard wooden surface, pushing the splitter down through the nodes.

Materials and tools for sanding and smoothing. Left to right: sanding block, sandpaper, steel wool, flat rasp, round rasp, shaver.

I bought a power drill and drywall screws. Although blatantly "un-Japanese," screws work quickly and efficiently to join bamboo to bamboo and bamboo to lumber. Screw heads can be covered by lengths of split bamboo or decorative ties. The best type of screws to use with bamboo are thin-diameter, with threads close to the head, a thin shank, and a Phillips head. Drywall screws, which range from 1 to 3 inches (2.5 to 7.6 cm) long, work well. For outdoor work, use galvanized decking screws. Their shafts are narrow and range in length from 1-½ to 3-½ inches (3.8 to 8.9 cm). Wood screws aren't recommended, because their shank broadens toward the head; this causes the bamboo to split if the pilot hole isn't as wide as the shank. If you want a screw head to be flush with a rounded culm, use a countersink drill bit to hollow out a shallow crater before you drive in the screw.

Smoothing

The edges of split bamboo can be surprisingly sharp, and the pain of a bamboo sliver under a fingernail isn't soon forgotten, so wear work gloves. For finer projects in which you handle the object, use a beveled knife or shaver to smooth the sharp edges of the inner wall. A few quick strokes, and splintery edges are gone! Additional smoothing with a sanding sponge makes the bamboo feel like silk. Use a curved or round rasp to smooth the edges of holes drilled for pegs or poles.

Joining and Attaching with Screws and Nails

I used only traditional Japanese ties of twisted black hemp to join my first bamboo projects. After hundreds of knots and worn-out hands and shoulders,

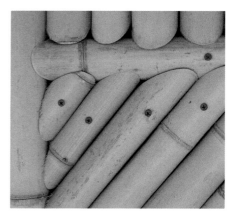

If you sink a screw directly into bamboo, the culm cracks and eventually splits. Unlike wood, bamboo doesn't "give." You can use screws successfully by first drilling pilot holes. The trick is to use a drill bit the same diameter as the screw, not smaller as you do with wood. When attaching bamboo to lumber,

drill a pilot hole completely through the two walls and just slightly into the wood. When you drive in the screw, secure it solidly in the lumber. When attaching bamboo to bamboo, rasp the areas where the poles intersect. With the two pieces held in place, drill a pilot hole through the first three walls, then follow with a screw. If the two poles wobble after joining, secure with wire or cordage.

As it enters the hollow internode, a screw point may not easily find its way to the next pilot hole. If it misses, it could create a crack. Prevent this by taking care to drill the pilot holes in a straight line. If your drill has a built-in level, use

it! It also helps to work with two drills—one fitted with the drill bit and the other with a Phillips-head bit—to eliminate the task of changing bits. It's also helpful to keep the poles very stable with your hands or quick-release clamps so they don't shift between the time you drill pilot holes and drive in the screw.

Nails penetrate bamboo easily but are difficult to remove. It's easier and less damaging to remove screws with a reversable power drill than to pull out nails. Nails do, however, work well when attaching bamboo to lumber. For a tight fit, used *ringed nails* (also called *panel nails*). The shafts have raised metal which grabs the fibers of bamboo. Be sure of the nails' placement beforehand; once in, they're difficult to remove. As with screws, you must first drill a pilot hole with a diameter equal to the nail.

Lashing and Binding

For thousands of years, lashing with thin strips of bamboo or twisted plant material has been the most universally embraced technique for joining bamboo. Lashing can be as simple as wrapping and tying two pieces together with cordage, or as complex as securing angled architectural joints that in turn support roof beams and joists. A few simple lashings are all you'll need to get started.

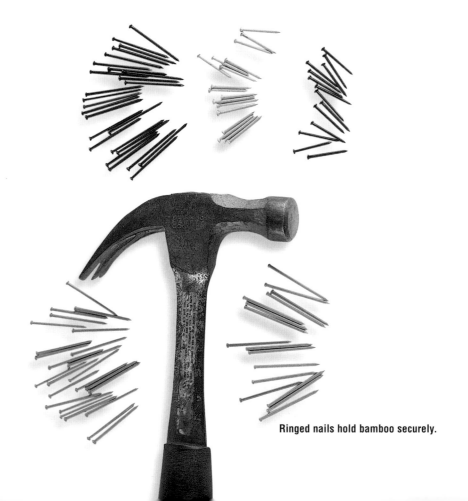

Ringed nails hold bamboo securely.

Top, left to right: square lashing, front crossed lashing, spiral front and back, lashing with spiral, lashing with twisted spiral. Bottom, left to right: traditional Japanese tie, square lashing, basic cross tie, square lashing, traditional Japanese tie.

CORDAGE

Most tying material is made from natural fibers. Coconut fiber, hemp, and sisal binder twine are suitable for outdoor projects. For decorative projects, waxed linen and twisted sea grass work well. Jute is flexible and easy to tie but rots quickly; the rougher strands of hemp and coconut better withstand the stress of sun and rain.

Traditionally, the Japanese use brown hemp twine and black hemp twine, which is made by dyeing the brown twine with carbon black. After several seasons of weathering, the rough fiber twine frays and fades to a silver gray. There is an abundance of alternative lashing materials, including fluorescent masonry cord, bicycle inner tubes, vines, and telephone wire. Creativity rules!

Clockwise, from upper left: dyed black hemp, dyed brown hemp, binder twine, twisted coconut fiber, dyed red hemp, twisted seagrass.

Here are several all-purpose ties, useful for basic attachments. If lashing intrigues you, apply more complex wrapping and knots.

BASIC CROSS TIE

This lashing steadies the two poles by crossing in back and tying in front. Start with 18 inches (45.7 cm) of cordage. Referring to figures 1 through 3, proceed as follows:

1. Fold the cord in half and place diagonally across the intersection, both ends pulled toward the back.

2. In the back, twist the ends tightly together.

3. Bring the two ends back to the front and tie with a square knot.

Figure 1

Figure 2 (back)

Figure 2 (front)

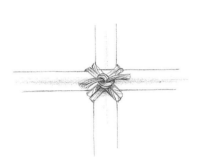

Figure 3

SQUARE LASHING

This is a sturdy lashing, good for structures. Starting with 36 inches (91.4 cm) of cord, refer to figures 1 through 3:

1. Attach one end with a clove hitch. Wrap the cord in a counterclockwise direction, bringing it under, over, under, and over the four poles. Pull tight at each wrap. Complete two or more full rotations.

2. Wrap between the poles, pulling tight and keeping each turn taut.

3. Tie the ends together in a square knot. Cut the ends.

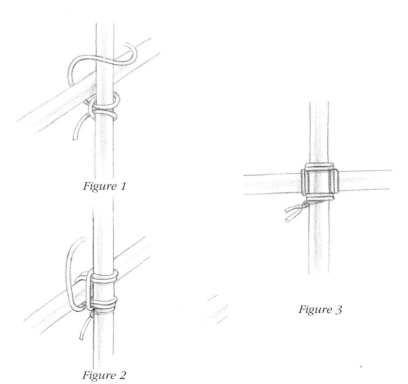

Figure 1

Figure 2

Figure 3

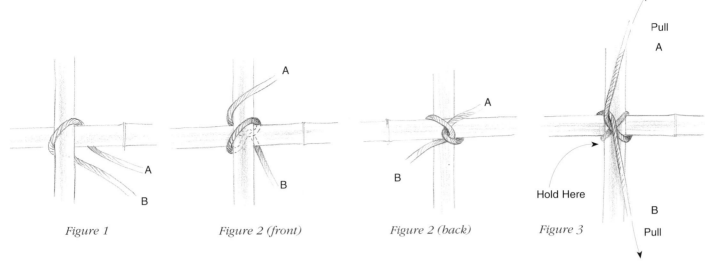

Figure 1   Figure 2 (front)   Figure 2 (back)   Figure 3

TRADITIONAL JAPANESE TIE

This tie is suited for use as a decorative knot, not a strong lashing. Starting with 37 inches (94 cm) of twine, and referring to figures 1 through 5:

1. Place the twine diagonally across the intersection with both ends toward the back. Adjust the twine so cord A over the pole is 4 inches (10.2 cm) longer than cord B.

2. In back, cross and tightly twist. Bring the ends forward, the longer one (A) from the upper left corner and the shorter one (B) from the lower right corner.

3. Cross and tighten at the intersection. Bring cord A up. Keep taut by pinching the twine.

4. Bring A over and under B then up, making a counter-clockwise loop. As it meets the intersection, place A under itself. Meanwhile, pull B up to keep the tie centered. Keep the knot tight by pinching, and

Figure 4

manipulate A to make the loop smaller.

5. Twist B under A. Continue to pinch, and take B through the loop.

6. Carefully pull and manipulate A and B in opposing directions to snug up the knot, working to get it as tight and close to the pole as possible.

7. Maneuver the loop to a diameter of 1 inch (2.5 cm) or less, and pinch it to close it. Hold in place.

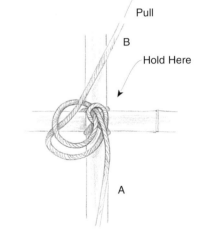

Figure 5

8. While keeping the intersecting twine in the center, take the twine in your right hand and move it clockwise under, around, and all the way through the loop.

9. While pinching with your left hand, use your right hand to pull the lower left twine down to complete and tighten the knot. Clip the ends to about 1 inch (2.5 cm).

USING WIRE

Galvanized steel wire and copper wire work well as lashings if the bamboo is cured and you use flexible wire and snub-nose or needle-nose pliers to achieve a tight twist. If green bamboo is lashed with wire, the joins will loosen when the bamboo dries and shrinks. If the wire isn't flexible enough to wrap tightly around a joint, the bamboo will slip and slide. To get a truly tight joint, rasp the points of intersection, join them with a screw, and stabilize the joint with wire. This works well during project assembly, when adjustments can be made before the final stage of wrapping. If desired, hide screw heads by twisting wire ends into a decorative spiral. Wire can also be lashed and twisted over cordage to stabilize structures.

FRONT CROSSED LASHING

1. Start with a piece of wire 18 inches (45.7 cm) long. Bend it

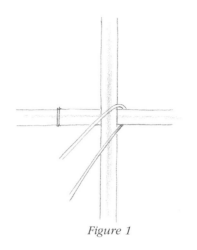

Figure 1

in half and place it behind the horizontal pole. Bring the lower end up and across the other end.

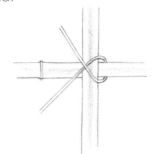

Figure 2 (front)

2. Make a sideways twist and take the ends to the back.

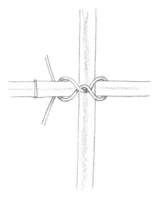

Figure 2 (back)

3. Use pliers to twist the ends together. Cut off the ends.

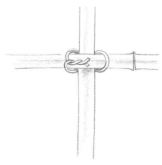

Figure 3

WIRE CROSS TIE WITH SPIRAL

Follow the directions for the Basic Cross Tie on page 44, and use needle-nose pliers to bend the ends into a spiral. If you're working with thin-gauge wire, twist the ends together then bend them into a spiral.

SPIRAL FRONT AND BACK

1. Start with a 6-inch (15.2 cm) piece of sturdy wire. Use nee-

dle-nose pliers to make a 90° bend in the middle. Starting at the bend, work the wire into a spiral.

2. Using a drill bit with a diameter that matches the wire, drill a hole through the poles in the center of the intersection.

3. Insert the straight end of the wire through the holes. In the front, adjust and flatten the spiral to fit the bamboo's curve. In

Bamboo joinery techniques were used to make this shade canopy. Design by Harry Abel

back, use needle-nose pliers to pull the straight end tight. Holding the front spiral steady, bend the back end into a spiral.

Joints

Simple joinery is done by inserting the end of a small-diameter bamboo pole into a hole drilled through one or both walls of a pole that's larger in diameter. Start by marking the outline of the smaller piece onto the larger. Select a Forstner or hole-saw bit just slightly smaller than the outline, and drill through the wall. Use a round rasp or Japanese knife to scrape the walls of the holes, customizing

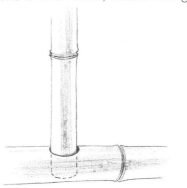

the fit to the incoming pole. You can also drill through the other wall and slide the smaller diameter pole through both walls.

MAKING PEGS

Pegs are useful for stabilizing joints. They're the equivalent of wooden nails and are made from the thick-walled base of a culm. To make pegs, first estimate the peg length required, a length equal to the diameter of the pole, plus 1 inch (2.5 cm) on either side. Cut the length from the culm base, split it in half, and use a hammer to knock out the diaphragms. Next, determine the diameter of the peg and make the splints slightly larger. Bracing one end on a work surface, use a single-bevel knife to shave the pulpy inner walls to a tapered end. Be sure to leave the hard

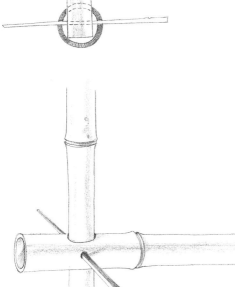

outer wall intact as it is the strongest part of the peg.

To peg a joint, drill a hole through all four walls. Place the peg in the hole so that the outer wall of the peg faces toward one side of the pole, rather than facing up or down. Otherwise the pole may split. Use a hammer to pound in the peg. The fibers of the peg compress between the drilled holes and fit snugly. For even more stabilization, cover the tapered peg with wood glue before driving in. Use a fine-tooth saw to cut off the protruding ends.

FISHMOUTH JOINTS

This term refers to the shape of the carved end of a bamboo pole; it resembles the open mouth of a fish. When placed against the curve of an adjoining pole, it fits closely against it as shown in figure 1 on page 48. To make the fishmouth, estimate how deep the cut-out curve needs to be in order to fit snugly against the adjoining pole. Mark this distance on the opposite side of the pole. With a fine-tooth saw, make short diagonal cuts to form a V, with the point of the V at the mark. To smooth the cuts, rub with a sanding pole made by wrapping a sheet of sandpaper around a pole that's the same diameter as the pole

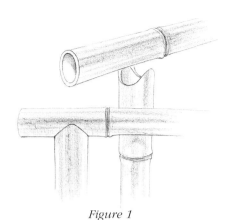

Figure 1

the fishmouth will join. Leave 6 inches (15.2 cm) unwrapped for a grip, and secure with masking tape. Fit the sanding end up against the cut, and rub the curved edges back and forth until the cut is smooth and fits well with the pole to be joined. As you work on a project, keep several poles wrapped with fine-, medium-, and coarse-grit sandpaper handy.

Bending Bamboo

Sometimes curved culms of bamboo will enhance a project's

structure or aesthetics. If you don't have naturally-curved lengths, two bending techniques work well. Both can also be used in reverse to take the curve out of a pole.

FIRE AND WATER

This technique creates gradual, slight curves. The heat of a fire is used to soften the oil and wax in the culm, the fibers are gradually stretched, and water is applied to cool the culm so it holds the bend. First, mark the section of the culm to be curved, and mark the midpoint of the inside curve. Choose a work area with two trees or stationary posts to serve as leverage to make the bend. Build a contained fire nearby and have ready a cloth or large sponge, a large bucket of cold water, and if possible, a friend to help you.

Rotate the area of the curve over the fire until it's hot (but don't scorch it). Quickly position

the hot culm between the trees or posts. While your partner stabilizes one end, firmly pull or push your end until you feel a slight give in the culm. Move the warm area back and forth against the post. When you feel the culm has bent to its maximum, stop and hold it firmly in position while your partner drenches the section with cold water. Repeat the process of heating, positioning, bending, and cooling, working alternately to the left and right of the midpoint, until the desired curve is achieved.

KERFS

To create a more extreme curve, you can cut triangular slices called *kerfs* along the inner bend of a culm as shown in figure 2. First, mark a line along the midpoint of the inside curve. Mark and saw out identically sized, equidistant triangular pieces. Using your own strength or working between

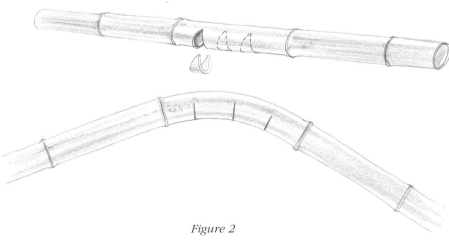

Figure 2

fixed posts, bend the culm. If the curve isn't round enough, cut out more kerfs or make the existing ones larger. Maintain the curve by tying or screwing the ends of the pole to other fixed poles.

Flattening

There are two ways to transform bamboo's clylindrical form into a relatively flat surface: one is a Japanese technique and the other, a Latin American technique. Either way, the idea is to break apart some—but not all—of the fibers to allow the culm to open up and spread out.

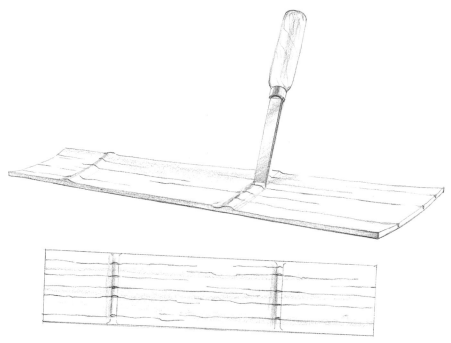

Figure 3

JAPANESE FLATTENING TECHNIQUE

Split a 2- to 3-inch (5.1 to 7.6 cm) culm in half, and use a hammer to knock out the diaphragms. Lay the culm inner side facing up and clamp it securely to a stable surface. With strong, sure strokes, hammer the chisel straight into the pulp of the diaphragm walls, penetrating and separating the cellulose fibers. Don't fully penetrate to the node, however. Repeat the chisel work at regular intervals along each node as shown in figure 3, with the distance between chisel strokes determined by the desired width of the finished slats. For example, if you want your slats to be 1 inch (2.5 cm) wide, separate your chisel marks by 1 inch or

more. Take care to strike each of the diaohram walls at slightly altered points across the length of the culm. Otherwise, the fibers may split perfectly along the grain and the culm may break apart. When you finish chiselling, turn the piece over and use the weight of your hands and body to gradually flatten the piece. If necessary, make additional chisel strokes to get a flatter surface. Lay pieces of lumber across it, and clamp it down overnight. Screw or nail in place.

ESTERILLA

Derived from the Latin American Spanish word meaning mat or matting, this technique flattens the culm by creating an accordion-like series of splits.

Bamboo esterilla is used in Central America as matting, wall paneling, and table veneers. Use the midsection of the culm, where the diameter is more or less constant.

Begin by splitting a pole in half, and use a hammer to knock out the diaphragms. From one side, measure 3/16 inch (4.8 mm); use a knife and mallet or the tapping technique to embed the blade. Twist the blade, moving it down to create a split about three-fourths of the way down. Flip to the other end, measure 3/8 inch (9.5 mm), and split along the grain the same distance as your first split. Flip again, measure 3/8 inch (9.5 mm) from the previous split, and split. Continue back and forth until you reach the other side.

Gently spread open the split pole as shown in figure 1. With a knife

SPLITS   OPENED

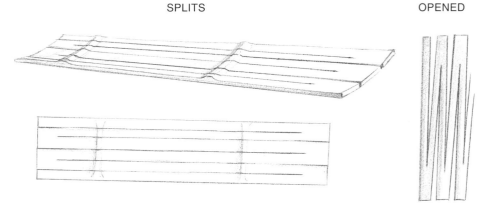

Figure 1

that has a long blade, cut off the diaphragms of the inner wall. Run the piece through a planer, shaving off the rough inner pith to achieve a flat, uniform surface. The esterilla is now ready for its intended purpose. Refer to the Side Table with Bamboo Esterilla on page 74 for one example.

Finishing and Preserving

Left untreated, bamboo weathers quickly when exposed to the elements. Finishing its outer surface preserves the structure while giving it richness and sheen.

SWEATING

The process of heating fresh or cured bamboo over a fire or flame is called *sweating*. Heat causes the oil and wax in the culm to rise to the surface and bead up like sweat. By removing excess secretions and rubbing with a cloth, you can give the outer layer a hard surface and beautiful sheen.

First, wash and scrub off any grime from the lengths you'll sweat, because dirt gums up the released oil and is difficult to remove. Make a fire in a safe, outdoor enclosure such as a barbecue grill or a galvanized metal tub with holes in its bottom and sides, set on bricks. The fire should burn hot and fast, so have a good supply of kindling and fuel nearby. If an outdoor fire isn't an option or you want to sweat just a few poles, you can use a stove or handheld propane torch. For large projects, a roofing torch that runs off a five-gallon (19 L) tank of propane works well. Scraps of bamboo make good fuel, but be careful to use only pieces that are completely split open. Lengths of bamboo with closed internodes can explode in fires and send fragments flying. One Malayan tale claims that such a loud and sudden noise is what gave the plant its name: "bam-BOO!"

Hold a portion of the pole 6 to 12 inches (15.2 to 30.5 cm) above the fire, moving it in sweeping circular motions. When you see wax beading up on the surface, let it build up, then remove the culm from the fire and use a soft cloth to quickly wipe off the oil and buff the surface. Return an overlapping section of the culm to the fire and repeat the rotating, wiping, and buffing process. The green culm will change to a shiny tan. Unless you want to create dark areas, be careful not to hold one area of the culm over the flame too long, or the oil will burn and scorch the surface.

POLISHING

A quick way to make a dull culm shine or to remove the chalky-white deposit on the surface of some bamboos is to use fine-grade steel wool. Wet the bamboo or dip the steel wool in water, then rub the pad up and down. Work in sections until the surface achieves a luster.

STAINS AND SEALERS

Exposed to the elements, the surface layer of bamboo eventually breaks down and becomes pitted. Mold and mildew may accumulate and lead to decay. On the good side, the rough surface will accept finishes which otherwise won't adhere to bamboo's waxy silicone surface.

Stains give cured bamboo a new lease on life. Wash cured and weathered bamboo to remove dirt and mildew, and let dry. Use a soft cloth to rub the

stain onto the culm, starting at the base and moving up. The stain adheres to the surface, rather than penetrating it, and a second coat may be necessary to achieve the full depth of color. A sealer should also be applied to outdoor projects that have become pitted. Suitable products include polyurethane and water sealer with ultraviolet protection, and varnish. For projects not exposed to the weather, tung oil, camellia oil, and floor wax work well. Apply several coats with a brush or cloth.

SCRAPING AND DYEING

If you want a refined, fresh-looking surface, you can scrape away the outer layer of green or cured bamboo to expose the delicate longitudinal strands of fiber. Hold the top of the culm at an angle with its base resting on a wooden surface. Hold the knife blade on an angle and move it up and down in a repetitive movement as the delicate shavings fall. Patience and strong arm muscles can give your bamboo a beautiful surface. Sealers and stains adhere to a scraped surface. Use wood dyes from hardware stores or fabric dyes formulated for cellulose fibers. Apply one or more coats, and finish with a coat of sealer.

ANNUAL MAINTENANCE

It's a good practice to make an annual inspection of outdoor

Gel stains are non-drip and work well on bamboo. "Golden oak" tones give a natural look.

bamboo structures. Fall is a natural time to do this, after vegetation has died down but the weather is still mild. Remove any vines, vegetation, and debris lodged around or between the poles. Wash the structure by thoroughly wetting it with a garden hose, then spraying with a nonsudsing cleanser or a highly diluted bleach-and-water solution. Give the cleanser time to act, then loosen the grime with a scrub brush, paying special attention to joints and intersections. Rinse thoroughly with fresh water. Let dry, then follow with stain or sealer.

Chapter 6

Bamboo in the Garden

Bamboo bridge by Kat Semrau. Photo by Rita Randolph

JUST ABOUT ANY GARDEN CAN BENEFIT FROM STRUCTURES MADE OF HARVESTED BAMBOO. UPRIGHT POLES WITH THEIR TOP ENDS CUT JUST ABOVE THE NODE MAKE FRIENDLY RESTING POSTS FOR THE BIRDS WHO CONTROL THE INSECT POPULATION. BAMBOO TRELLISES GIVE GRACEFUL SUPPORT TO FLOWER-ING AND FRUITING VINES.

IF YOU GROW YOUR own bamboo, the garden is a good place to use your harvested culms. Keep assorted lengths stored near your garden; they'll come in handy for all sorts of garden projects and puttering.

Stabilizing Poles

You'll find several techniques handy for stabilizing freestanding bamboo structures. To set poles in the ground, cut the bottom end on the diagonal. Use a hammer to pound a piece of rebar through the base 18 to 24 inches (45.7 to 61 cm) into the pole. Knock out the nodes and ream out as much membrane as possible. This will give the dirt a place to go when the pole is hammered into the ground. Use a rubber mallet to drive in the pole; if you use a hammer with a steel head, place a wood scrap over the top of the pole to protect it.

You can also use rebar fitted with bamboo sleeves. This technique keeps the poles off the ground, lengthens the life of the structure, and allows you to remove it for storage or repair. Use a hammer to pound rebar 3 to 5 feet (0.9 to 1.5 m) through the pole, knocking out the diaphragms. This creates a hollow tube or sleeve. Drive a 3- to 5-foot (0.9 to 1.5 m) length of rebar 18 to 24 inches (45.7 to 61 cm) into the ground or deep enough to assure stability. Then slide the hollowed bamboo pole over the rebar. To decrease rot, place bricks on either side of the rebar to keep the bamboo off the ground.

Clockwise from upper left:
Bamboo trellis with copper tubing. DESIGN AND PHOTO BY PETER GALLAGHER.

This bamboo trellis supports flowering perennials.

A length of bamboo is combined with straw to create a birdhouse.

Bamboo creates enchanting vignettes when used as supports for bushes and shrubs.

A bamboo pipe and dripper add calming sounds to a water garden.

53

Lightning Trellis

THANKS TO THE ZIGZAG SHAPE OF GOLDEN BAMBOO (*PHYLLOSTACHYS AUREA*), THE HORIZONTAL POLES OF THIS DELIGHTFUL TRELLIS REST IN THE CROOKS OF THE VERTICAL POLES. THE OVER-AND-UNDER PATTERN GIVES DIMENSION AND ADDITIONAL VISUAL INTEREST. ARRANGE AND ADJUST THE FULL-LENGTH POLES ON A FLAT SURFACE TO GET THE BEST FIT BEFORE YOU CUT.

Instructions

1 Refer to tools and materials listed on page 56. Select one of the nine vertical poles to serve as the center pole. It should be strong and have an appealing zigzag shape. Use the hammer and rebar to knock out the lower 3 feet of diaphragms.

2 On a large, flat surface, lay out the vertical poles in a fan shape with the zigzag ends up. Start by placing the center pole in the middle and work outward, bringing the poles close together at the base, and fanning them out to 60 inches at the widest part. Position and adjust the poles to get a good fit.

3 One by one, lay the vertical poles over and under the horizontals where poles and zigzags fit together. Work to create a surface with dimension. See figures 1 and 2.

4 When you're satisfied with your layout, use the saw to cut the poles to their finished lengths.

5 Where the poles intersect, drill pilot holes and attach with the drywall screws on one side, then turn over the structure and screw from that side.

6 At the installation site, mark the center point, and hammer in the rebar 18 to 24 inches deep. Place a brick on each side of the rebar.

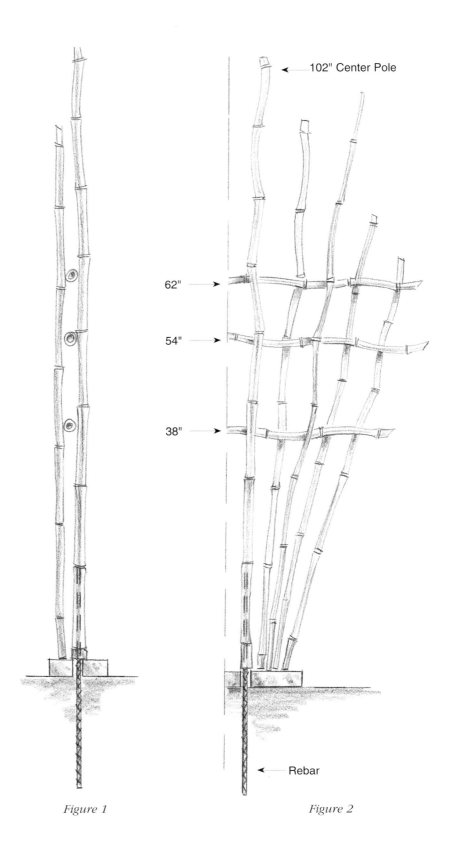

102" Center Pole

62"

54"

38"

Rebar

Figure 1

Figure 2

Materials

5 1¼-inch drywall screws

9 1⅝-inch drywall screws

9 2¼-inch galvanized decking screws

Rebar, 4 feet long, to serve as center pole support

2 bricks

Tools and Supplies

Fine-tooth saw

Power drill and assorted drill bits

Sanding sponge

Wire (optional)

Nails (optional)

Cutting List

Description	Qty.	Material	Dimensions
Verticals	9	Golden bamboo poles, zigzag-shape	1" to 1¾" x 6' to 8'6"
Horizontals	3	Golden bamboo poles, zigzag shape	1¼" x 38", 54", and 62" respectively

Metric Equivalents

1"	2.5 cm	1¾"	4.4 cm	38"	96.5 cm	3'	0.9 m
1¼"	3.2 cm	2¼"	5.7 cm	54"	137.2 cm	4'	1.2 m
1⅝"	4.1 cm	18"	45.7 cm	60"	1.5 m	6'	1.8 m
		24"	61 cm	62"	57.5 cm	8'6"	2.5 m

7 Hold the trellis over the rebar and slide the central pole over the rebar, resting the base of the poles on the bricks. If more stability is required, use the wire and nails to attach the trellis to the wall.

Water Dipper

with assistance from Doug Lingen and Reed Hamilton

B AMBOO WATER DIPPERS ARE PLACED AT THE ENTRANCE OF MANY JAPANESE TEMPLES AS AN INVITATION FOR VISITORS TO TAKE PART IN RITUAL CLEANSING. THE DESIGN OF THIS DIPPER IS APPEALINGLY SIMPLE AND FUNCTIONAL AND CAN BE CRAFTED FROM SMALLER-DIAMETER BAMBOO AS WELL. TO GIVE IT A LONGER LIFE, REST THE DIPPER UPSIDE DOWN TO THOROUGHLY DRAIN THE WATER WHEN IT'S NOT IN USE.

Materials

1 bamboo splint, 21 inches long, 1¼ inches wide, ¼-inch thick, to serve as the handle

1 piece of bamboo, 4 to 5 inches long, 3½ inches in diameter, to serve as the bowl*

*This piece should have 1 node and ⅜-inch thick walls

Metric Equivalents

¹³⁄₆₄"	5.2 mm	1⅛"	2.8 cm
¼"	6 mm	1¼"	3.2 cm
½"	1.3 cm	2"	5.1 cm
⅜"	9.5 mm	3½"	8.9 cm
⅝"	1.6 cm	4"	10.2 cm
¾"	1.9 cm	5"	12.7 cm
¹³⁄₁₆"	5.2 mm	7"	17.8 cm
1"	2.5 cm	21"	53.3 cm

Tools and Supplies

Quick-release clamps

Power drill (or drill press) with 1-⅛-inch Forstner drill bit and ¹³⁄₆₄-inch drill bit

Fine-tip permanent marker

Ruler or straightedge

Fine-tooth saw

Japanese cutting knife

Shaver

Sanding sponge

Sandpaper, medium grit

Pure beeswax candle

Matches or lighter

Piece of flexible, waxed cardboard, 1 x 2 inches

Instructions

1 To make the handle, mount the Forstner bit in the drill or drill press. Hold the bit to one end of the splint and create a curve on the end. This is most easily accomplished with a drill press, but if you use a handheld drill, use a vise or clamp the piece securely to your work surface, and hold the drill steady as you drive it through the splint.

2 Starting at the curved end of the splint, mark a paddle-shape outline that's ¾ inch wide at one end and narrows to a ½-inch neck. Along each side of the handle, use the ruler to make straight lines that start out ½ inch apart at the neck, then very gradually widen so that at a point 7 inches up from the neck, they are ⅝ inch apart. Maintain the ⅝-inch width to the end of the handle.

3 Referring to figure 1, use the knife to carve out the base, neck, and sides of the handle along the lines. Use the knife or shaver to remove splintery edges,

and smooth with the sanding sponge.

4 Now you'll make the bowl. From the length of bamboo 3-½ inches in diameter, use the fine-tooth saw to cut just below the node. Make sure you cut evenly across so the bowl sits level. If you have doubts about making an even cut, mark a straight line around the perimeter to guide you as you cut.

5 Use the knife to round off the inner and outer rim. Follow with the sanding sponge, smoothing the rim of the base.

6 As shown in figure 2, decide the side of the bowl to which you wish to attach the handle. Trace the curved outline of the handle base onto this area, 1 inch below the rim.

7 Use a vise or clamp the bowl so it sits open end up. Holding the $^{13}/_{16}$-inch drill bit at the angle desired for the handle, drill in between the outline, the bit pointed down and in. Drill a series of holes overlapping each other along the curve of the outline until the opening has been made.

8 Use the knife to cut out the pieces of bamboo that remain between the holes. Then fold up a piece of the sandpaper and run it back and forth through the opening until the inside edges are even.

9 Insert the base of the handle in the hole. The joint should be snug with the handle fitting securely in the bowl. If the handle won't quite go in, mark and rework until the fit is satisfactory. You can also trim the outer wall of the handle to decrease its width.

10 Now you'll attach the handle to the bowl. From the inside of the bowl, hold the

waxed cardboard securely over the opening. Now light the beeswax candle. Let the hot wax drip into the curved opening until it's half full. Blow out the candle and quickly insert the handle into the bowl. Hold the cardboard tight so the wax doesn't ooze into the bowl and hold the handle and bowl steady for several minutes until the wax cools. Place in a cool place until the wax thoroughly hardens.

11 Use the knife to remove excess wax from the inside or the outside of the bowl.

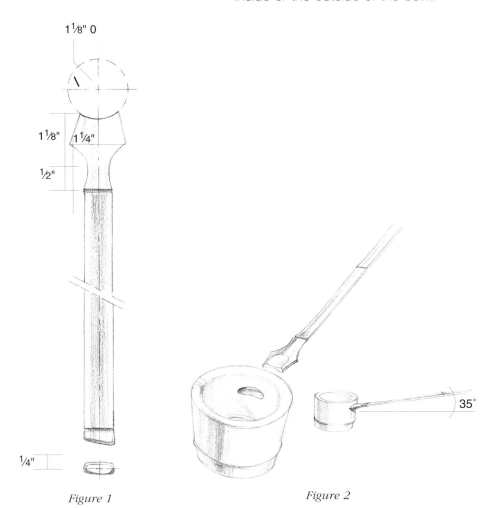

1⅛" 0

1⅛" 1¼"

½"

¼"

Figure 1

35°

Figure 2

Low Garden Trellis

Designer, Anita Matos

THIS SEE-THROUGH, DIAMOND-PATTERN TRELLIS IS A NATURAL FOR
CUCUMBERS, GOURDS, AND OTHER CLIMBING VINES. AS SHOWN HERE, IT
CAN UNIFY A GARDEN GONE GLORIOUSLY WILD WITH CULTIVATED AND
"VOLUNTEER" FLOWERS AND VEGETABLES. A THIRD POST IN THE TRELLIS PROVIDES EXTRA
SUPPORT FOR WEIGHTY PRODUCE.

Materials

2 lengths of rebar, each 4 feet long

60 drywall screws in lengths of 1, 1¼, 1⅝, and 2 inches

Galvanized wire, 16-gauge

Copper wire

Binder twine

Tools and Supplies

Hammer

Measuring tape

Power drill and assorted drill bits

Pliers

Wire cutters

Cutting List

Description	Qty.	Material	Dimensions
Horizontals	3	Bamboo poles	1½" x 7'
Posts	2	Bamboo poles	1½" x 4'
Long diagonals	11	Bamboo poles	½" to 1" x 6'
Short diagonals	3	Bamboo poles	½" to 1" x 4'

Instructions

1 Working with the two 4-foot poles that will serve as posts, use the hammer and a piece of the rebar to knock out the lower 3 feet of diaphragms.

2 On a large, flat surface, lay out the posts 6 feet apart. Place the horizontal stringers on top, one 3 inches below the top of the post, one 3 inches above the base, and one in the middle. Drill pilot holes at the intersections of posts and stringers, and drive in the 2-inch screws.

3 Mark the center of each horizontal. You'll use the marks as reference points when placing the diagonals to form the diamond pattern.

4 Place five of the long diagonals and one of the short diagonals on the frame, all leaning in the same direction. Center the diagonals, drill pilot holes, and use the appropriate screws to secure the ends to the frame.

5 Turn over the trellis assembly. Lay out the remaining diagonal poles in the opposite direction. Voilá, diamond patterns! Adjust, drill pilot holes, and screw at intersections.

6 To install the trellis, mark the placement of the two end posts. Use the hammer to pound the rebar into the ground at those points. Lift up the trellis and slide the posts over the rebar.

7 Use the wire or cordage to reinforce intersections that need extra support.

Metric Equivalents

½"	1.3 cm
1"	2.5 cm
1¼"	3.2 cm
1½"	3.8 cm
1⅝"	4.1 cm
2"	5.1 cm
3"	7.6 cm
3'	0.9 m
4'	1.2 m
6'	1.8 m
7'	2.1 m

Plant Stakes and Markers

RATHER THAN HUNTING FOR FALLEN BRANCHES TO MARK SEEDLINGS OR TIE UP LEANING PLANTS, YOU CAN EASILY SPLIT BAMBOO INTO HANDY STAKES AND MARKERS. THEY'RE AN EXCELLENT USE FOR THOSE IRREGULAR SCRAPS, AND ARE LONG-LASTING. IN THE FALL, PULL THEM OUT OF THE GROUND, SHAKE OFF THE DIRT, AND STORE FOR USE IN THE SPRING. BUNDLES OF SMOOTH STAKES TIED WITH TWINE MAKE GREAT GIFTS FOR YOUR GARDENING FRIENDS!

Materials, Tools, and Supplies

5 or more bamboo poles, 12 to 60 inches long, 1¼ inches or greater in diameter

Splitting knife and mallet, or a four-way splitter

Hammer

Knife or shaver

Sanding sponge

Fine-tip permanent markers (optional)

Metric Equivalents

1¼"	3.2 cm
12"	30.5 cm
60"	1.5 m

Instructions

1 Use the splitting knife and mallet to split the lengths of bamboo in half, then hammer out the diaphrams. Split each length in half again, forming a total of four splits per pole. If you're using a four-way splitter, split the lengths, then remove the quartered node membranes with a knife.

2 Use the shaver or knife to smooth the edges and eliminate splinters.

3 Rub the sanding sponge up and down the splint several times to smooth the edges and the inner wall.

4 If desired, use the fine-tip permanent marker to write plant identifications or other information directly on the bamboo. Color them, too, if you're feeling festive!

Dripper and Deer Scare

Designer, Harry Abel

THIS PAIR OF WATER GARDEN ACCESSORIES WORKS TOGETHER TO CREATE A SIMPLE AND ENCHANTING DISPLAY OF MOVEMENT AND SOUND. A SLIGHT STREAM OF WATER FLOWS FROM THE *SOZU*, OR DRIPPER, INTO THE WAITING END OF THE *SHISHI ODOSHI,* THE DEER SCARE. THE SCARE'S HOLLOW MOUTH SLOWLY FILLS. AFTER THE WEIGHT OF THE WATER CAUSES THE SCARE TO TIP AND EMPTY, THE BASE END DROPS QUICKLY TO STRIKE A ROCK, PRODUCING A HOLLOW CLACKING SOUND. IN THE JAPANESE CULTURE, THE SLOW MOVEMENT OF WATER, THE ACTIONS OF FILLING AND EMPTYING, AND THE PUNCTUATION OF SUDDEN SOUND ARE COUNTERPOINTS TO THE STILLNESS OF THE GARDEN AND MARKERS OF THE PASSAGE OF TIME.

Materials

1 36-inch piece of ¼-inch rebar, to serve as support for dripper uptake pole

2 24-inch pieces of ½-inch rebar, to serve as supports for scare posts

Polyethylene tubing, 72 inches long, ½-inch outside diameter

1 smooth, curved, dark rock to serve as the sounding rock

Water source such as a pond or basin

Recirculating water pump

Tools and Supplies

Measuring tape

Fine tooth saw

Vise or clamps

Power drill with 1½ inch Forstner bit and ½-inch brad point bit

Hammer

Sanding sponge

Garden hose or spigot

Cutting List

Description	Qty.	Material	Dimensions
Dripper uptake pole	1	Bamboo pole	1½" x 36"
Dripper head	1	Bamboo pole with nodes at each end*	5" to 6" X 8½"
Dripper spout	1	Bamboo pole	1½" x 12"
Scare water collector	1	Bamboo pole with 4 or 5 nodes	3" x 36"
Scare pivot bar	1	Bamboo pole	½" x 6"
Pivot bar posts	2	Bamboo lengths with branches**	1½" x 24"

*This piece can be taken from the base of a culm at a point where the nodes are close and the culm diameter is widest.

**Cut the top of these pieces just above the node, and trim the two branches that extend from the node to 6 inches.

Instructions

Making the Dripper

1 Use the rebar and mallet to knock out all the diaphragms of the 36-inch uptake pole. Ream as close to the walls as possible to allow room for both the plastic tube and the 24-inch length of rebar. Determine which end of the uptake pole is closest to 1-½ inches in diameter, and use the saw to cut that end to a 45° angle.

2 Secure the head of the dripper in a vise or clamp it to a worktable. Use the Forstner bit to drill a hole through one wall of the dripper midway between the two nodes. Turn the head 90°, and drill another hole midway between the nodes. The holes will receive the uptake pole and the spout.

3 Now you'll make the spout. Take the 12-inch length of bamboo and determine which end fits most tightly into the head. Use the saw to cut this end to a 45° angle with the opening facing down. Cut the mouth end at a 45° angle with the opening facing up.

4 Determine the placement of the dripper in relationship to the deer scare. At the site intended for the dripper uptake pole, hammer the ¼-inch rebar 18 inches into the ground.

5 Refer to figure 1. Snake the plastic tubing through the base of the uptake pole until it emerges from the angle-cut top.

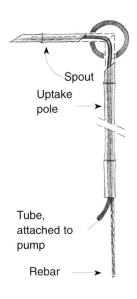

Figure 1

Maneuver the tube through the holes in the head, then fix the head on the pole. Slide the end of the tubing through the spout, stopping before it is visible. Secure the angled end of the spout into the head, making sure the angled cuts fit together and the tube is not crimped.

6 Insert the other end of the plastic tube into the water pump.

Making the Deer Scare

1 First, you'll make the water container. Select the base end of the 36-inch-long, 3-inch-wide pole. This will strike the sounding rock, and the opposing end will collect water. Use the saw to cut the base straight across below the node, and cut the mouth end at a 45° angle. Smooth the edges with the sanding sponge.

2 Use your thumb and forefinger to lightly grasp the pole at the center of the middle node. Drip water from the garden hose into the mouth of the scare. When the mouth is full, the tube should tip in your fingers, spilling the water. If it doesn't tip, test for the pivot point by sliding your fingers up and down the length. Mark this pivot point on both sides of the pole.

3 Use the brad-tip bit to drill through the mark on one wall into the mark on the other wall. Just as the point emerges, stop. Finish drilling the hole from the outside wall to prevent splitting and feathering. Sand the edges. Insert the wide pivot bar through the two holes and center it.

Installing the Deer Scare and Dripper

1 Use the hammer and rebar to knock out the diaphragms from the base up to, but not including, the top node of the two 24-inch post lengths.

2 Referring to figure 2, determine the placement of the scare in relation to the dripper. Place the two 24-inch pieces of rebar 5 inches apart, and hammer each one 12 inches into the ground.

3 Slip the posts over the rebar, aligning the base of the branches so they're equal in height.

4 Place the scare and pivot bar between the Y of the branches so the pole rocks easily back and forth. Trim the branches to a pleasing form.

5 Place the sounding rock so that when the water empties and the pole swings to rebalance, the returning end strikes the rock squarely. If necessary, adjust the water pressure and the place-

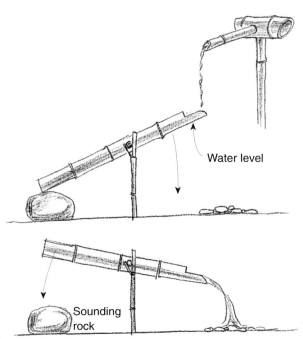

Water level

Sounding rock

Figure 2

Metric Equivalents

¼"	6 mm	8½"	21.6 cm
½"	1.3 cm	12"	30.5 cm
1½"	3.8 cm	18"	45.7 cm
3"	7.6 cm	24"	61 cm
5"	12.7 cm	36"	91.4 cm
6"	15.2 cm	72"	1.8 m

Chapter 7

Furnishings and Accessories

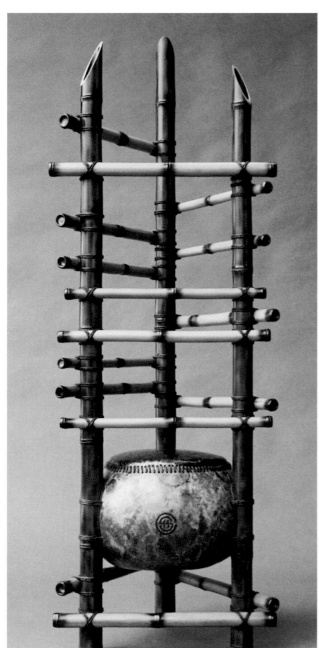

Cal Hashimoto, *Spirit Vessel* - Series V, 1997, 44 x 15 x 15 in. (1.1 m x 38 x 38 cm); bamboo, gourd, brass, twine.
PHOTO BY CAL HASHIMOTO

Right: Cal Hashimoto, *Call to Center* - Series II, 1998, 90 x 26 x 5 in. (2.3 m x 66 x 12.7 cm); bamboo, brass gong, twine.
PHOTO BY CAL HASHIMOTO

BAMBOO IS AS FUNCTIONAL AND PLEASING INSIDE THE HOUSE AS IT IS OUTSIDE, AND HANDCRAFTED BAMBOO OBJECTS LEND A NATURAL, OUTDOOR FEELING. ITS STRENGTH AND SMOOTH, CYLINDRICAL SURFACE MAKES IT A VERSATILE MATERIAL WITH WHICH TO CRAFT TABLES, SCREENS, BENCHES, AND OTHER FURNISHINGS. IF YOU COMBINE BAMBOO WITH MILLED LUMBER OR OTHER MATERIALS, THE TEXTURAL CONTRAST ADDS RICHNESS AND DIMENSION TO THE FINAL PRODUCT.

IN ADDITION TO FURNITURE, it's easy to make simple bamboo objects to add a serene feeling to any room. Bamboo vases hung on walls or grouped on a tabletop make charming displays for hand-picked flowers. If you're entertaining guests, you can offer food in sushi trays and drinks in bamboo cups. Best of all, a few hand tools and bamboo poles are all you need to make these lovely accessories.

Clockwise from top: Bamboo screens are framed with Western red cedar.
DESIGN AND PHOTO BY GORDON POWELL

Bamboo torchiére lamp designed by John Scheer.

A sunroom at the Bamboo Farm in Savannah, Georgia is faced with split bamboo.

Barbara Schindler, *Harmony Hanger*, 2000, 24 x 24 x 2 in. (61 x 61 x 5 cm); bamboo, hemp twine, beads, amber and turquoise stones.

Cedar and Bamboo Inlay Table

with the assistance of Randall Ray

C RAFTSPEOPLE KNOW THAT ELEGANCE IS OFTEN ABOUT RESTRAINT, AND CHOOSE TO HIGHLIGHT ONLY ONE OR TWO DECORATIVE FEATURES IN THEIR WORK. THIS TABLE IS CONSTRUCTED OF BEAUTIFULLY FINISHED CEDAR AND MITERED INLAY MADE FROM BLACK BAMBOO (*PHYLLOSTACYS NIGRA*).

Tools and Supplies

Tape measure

Miter saw

Power drill with ⅜-inch and ¹⁄₁₆-inch bits, and ⅜-inch router bit

30 galvanized decking screws, 2 inches long

Paste construction adhesive

Circular saw or table saw

Wood glue

Fine sandpaper

28 3-inch wood screws

30 2-inch wood screws

12 tapered wood plugs

2 paintbrushes

Clear wood sealer

Wood stain, brownish black color

Fine-tooth saw

Splitting knife

Hammer

56 1⅝-inch ringed nails

Cutting List

Description	Qty.	Material	Dimensions
Frame	1	Western red cedar 4 x 4	8'
Frame and legs	1	Western red cedar 4 x 4	10'
Cleats	-	Pressure-treated 1 x 2 lumber	13' total
Tabletop support	1	Pressure-treated plywood	¾" x 18" x 24"
Inlay	10	Black bamboo poles	1½" x 4' to 5'

Metric Equivalents

¹⁄₁₆"	1.6 mm	16"	40.6 cm
³⁄₁₆"	4.8 mm	17"	43.2 cm
¾"	1.9 cm	18"	45.7 cm
⅜"	9.5 mm	24"	61 cm
1½"	3.8 cm	37⅝"	95.6 cm
1¾"	4.4 cm	41¼"	104.7 cm
1⅝"	4.1 cm	48"	121.9 cm
2"	5.1 cm	4'	1.2 m
3"	7.6 cm	5'	1.5 m
8"	20.3 cm	8'	2.4 m
13¾"	34.9 cm	10'	3 m
14"	35.6 cm	13'	3.9 m
15¾"	40 cm		

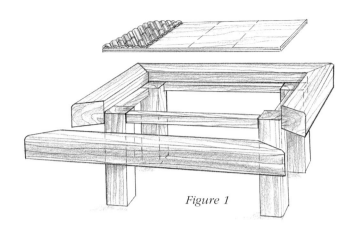

Figure 1

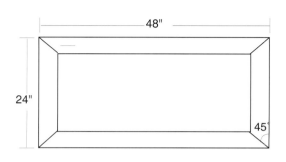

Figure 2

Instructions

Making the Table Frame

1 Refer to figures 1 and 2. Cut the 8-foot cedar 4 x 4 in half at a 45° mitered angle, each with a 48-inch long point. From the 10-foot cedar 4 x 4, cut two lengths at a 45° angle, each with a 24-inch long point.

2 Assemble the cedar pieces on the work surface, and use the ⅜-inch drill bit to predrill three ¾-inch-deep pilot holes in each corner (see fig. 3). Use the 1⁄16-inch bit to drill 2 inches into the center of each hole.

3 Dust off the shavings. Apply the adhesive to the ends of the cut pieces, assemble, and screw together with the 2-inch screws, countersinking ⅜ inch. Allow the glue to dry.

4 From the remaining 4 x 4 pieces, square-cut four legs, each 15¾ inches long. Rabbet the top end of each leg, as shown in figure 4. Cut and remove the three sections, leaving a tenon that fits into the corners of the frame.

5 Place the frame face down. As shown in figure 5, apply the adhesive and set the tenons of the legs into the inside corners of the frame. Drill a 1⁄16-inch pilot hole 2 inches deep. Drive in the 3-inch screws. Let dry. Sand or chisel off the excess glue.

6 Lay the table on its side, and pencil a line 1½ inches below the top edge of the frame. Of the 1 x 2 boards, cut two pieces 37⅝ inches long, and two pieces 13¾ inches long.

7 Predrill 3⁄16-inch pilot holes every 8 inches along the line you've drawn. Referring to figure 5, position the cleats you cut in step 6 between the legs, along the 1½-inch depth line. Screw in place with the 3-inch wood screws.

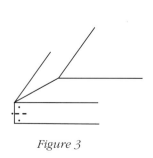

Figure 3

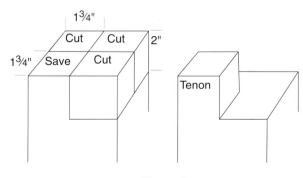

Figure 4

8 Place the table upside down. Cut three 2 x 2 boards to a 14-inch length. Referring to figure 6, set these stretchers across the width of the table, between the cleats. Drill two 1/16-inch pilot holes on an angle, then use the 2-inch screws to toenail the ends of the stretchers onto the cleats.

9 Place the table upright. Measure and cut the 3/4-inch plywood to 17 x 41 1/4 inches. After applying the wood glue to the top edge of the cleats, center the plywood and drop it into the table. Secure the plywood to the cleats with the 2-inch wood screws, six along the long sides, and three along the short sides.

10 Use the 3/8-inch round-over bit in the drill to router the outside top edge of the frame and the base of the legs, achieving a pleasing curve. Fill the screw holes with the wood plugs.

11 Clean off any dust or shavings, and use the paintbrush to apply two coats of the clear wood sealer to the cedar. Coat the plywood with the stain. Let dry.

Making the Tabletop Inlay

1 Brush the bamboo with the stain and let dry.

2 Use the tape measure and chalk to divide the plywood seat into four equal sections as shown in figure 7.

3 Use the fine-tooth saw to cut the bamboo into 10 16-inch lengths. Split the bamboo in half lengthwise, creating 20 pieces. Arrange the halves on the seat in a herringbone pattern as shown in figure 8, adjusting them to get a close fit and an even surface. Cut and split the remaining bamboo, and use it to complete the surface pattern.

4 Use the handsaw and miter box to cut the bamboo at a 45° angle. Work carefully and precisely to achieve a good fit.

5 Lightly sand the ends of the bamboo, and brush the surface and ends with the stain. Allow to dry, then lay the pieces back into the table. Drill a pilot hole through the end of each bamboo piece and into the plywood base. Hammer a ringed nail through the holes. Repeat with each piece until the inlay is complete.

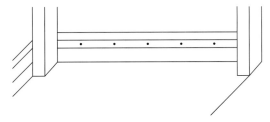

Figure 5

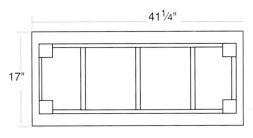

Figure 6

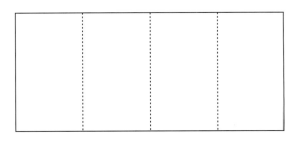

Figure 7

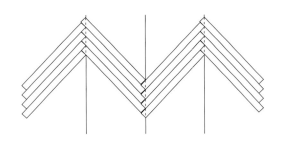

Figure 8

Side Table with Bamboo Esterilla

Designer, Francisco Plaza

ESTERILLA, A LATIN-AMERICAN SPANISH IDIOM FOR MAT OR MATTING, GIVES TEXTURE AND GRAPHIC INTEREST TO THE TOP OF THIS TABLE, A CIRCLE WITHIN A TRIANGLE. PEGS AND TRADITIONAL JOINERY TECHNIQUES ADD TO THE FINE CRAFTSMANSHIP OF THE TABLE.

Materials

16 drywall screws, 1-¼ inches long

Furniture tacks, ½ inch long

Wood glue

Brad nails, ½ inch long

Wood stain in brownish black color

Tools and Supplies

Source of flame

Straightedge ruler, 36 inches long

Pencil

Jigsaw

Circular saw

Power drill with: 1⅛ inch hole-saw bit; ¼-inch-diameter brad-point bit 10 inches long; and bit with a diameter that matches the furniture tacks

Fine-tooth saw

Belt sander with sandpaper, #40 grit

3 bungee cords

Miter box

Round wood rasp

Japanese knife

Hammer

Piece of string, 14 inches long

Splitting knife with long blade

Mallet with plastic or wooden head

Wood plane

Clean cloth

Paintbrush

Sanding poles (see page 46) with medium and coarse sandpaper

Cutting List

Description	Qty.	Material	Dimensions
Legs	3	Bamboo lengths	3½" x 20"
Rails	6	Bamboo lengths	1¼" x 21"
Tension braces	6	Bamboo lengths	1" to 1¼" x 16"
Esterilla	3	Bamboo lengths*	3½" x 20"
Splint for tabletop trim	1	Bamboo length** 2" x 6'	
Tabletop and leg attachments	6	Bamboo pegs	Diameter tapered from ¼" to ⅜", 6½" long
Rail and brace attachments	12	Bamboo pegs	Diameter tapered from ¼" to ⅜", 4" long
Jig	2	Scrap plywood	¾" x 24" x 24"
Tabletop base	1	Plywood	1" x 22" x 22"
Jig supports	6	2 x 2 lumber	12" long
Flattening strips	5	Scrap wood	¼" x 1" x 22"

*with a consistent wall thickness, taken from the midsection of the culm

**taken from upper midsection of a culm

Instructions

Preparing the Bamboo

1 Sweat and scorch the bamboo to produce a rich, brown sheen (see page 50).

Making the Jig

1 Referring to figure 1, use the ruler and pencil to draw an equilateral triangle with 16-inch sides onto one piece of scrap plywood.

2 Place the 20-inch legs top end down at the points of the plywood triangle, and trace around the legs. Use the ruler to draw lines connecting the outsides of the circles.

3 Use the circular saw to cut the straight edges of the plywood, and the jigsaw to cut out the space for the legs. Transfer this shape onto the other piece of scrap plywood, and cut it out.

4 Arrange the pieces of 2 x 2 lumber between the plywood forms as shown in figure 2, placing them toward the center so they won't get in the way of the legs.

5 Use the power drill and screws to attach the supports to the top and bottom plywood forms.

Making the Legs and Table

1 Use the fine-tooth saw to cut each 20-inch leg length just above the top nodes, straight across, being careful not to feather the outer wall. From the top, measure down 18 inches and cut straight across so the legs will sit flat on the floor. Use the belt sander to smooth the ends.

2 Place the legs in the jig, and wrap the bungee cords around them to keep them in place. Place the tabletop onto the plywood surface, setting it between the tops of the legs. When centered, there should be a 5/8-inch space between the legs and the edges of the tabletop to account for the trim that you'll add later.

3 Refer to figure 3. With the table legs still held in place by the bungee cords, measure and mark 1 inch down from the top of each leg. This indicates the top of the rails that will hold the tabletop. Then starting from the base of the legs, measure up 4 inches to mark the bottom of the rails.

4 Use the miter box and fine-tooth saw to cut a 60° angle at each end of the 21-inch rails.

5 Place the rails between the legs and check for fit. Use the

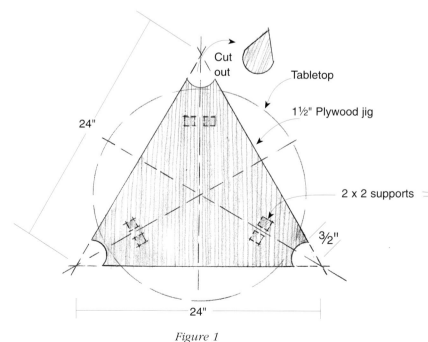

Figure 1

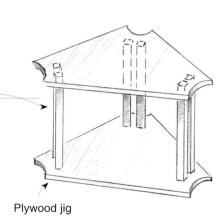

Figure 2

marker to outline the circumference of the rails onto the legs. At the top of the legs, outline each rail just below the mark. At the botttom of the legs, outline each support just above the mark.

6 Remove the bungee cords and the legs from the jig. Clamp each leg to the worktable, and use the drill with the 1⅛-inch hole-saw bit to make the holes in the legs. Take care to drill at the proper angle with the top and bottom holes in alignment.

7 Now you'll prepare the legs to be joined to the rails. Use the Japanese knife and round rasp to scrape the sides of each hole, customizing its shape to accommodate the end of the incoming rail and achieve a good fit. Reassemble the legs around the jig, insert the rails, and secure with the bungee cords.

8 Peg together the legs and rails as shown in figure 4. Use the 10-inch brad-point bit to drill through the leg walls and the angled ends of the rails. Use the hammer to drive a 6½-inch peg through the hole until it's tight, being sure to keep the outer wall of the peg facing the wall of the leg. Use the Japanese knife or saw to cut off the protruding peg ends. Repeat, pegging all the rails.

9 Now you'll cut and install the six 16-inch braces to give the table rigidity by putting tension on any loosely fitting pieces. Use the fine-tooth saw to make a fish-mouth cut at one end of each brace (see page 47), making the curve open enough to fit snugly against the leg. Repeat with all of the braces.

10 Repeat the pegging process outlined in step 8

to attach the braces to the legs. Use the brad-point bit to make a hole at an angle from the top of the brace down into the leg. Hammer in the 4-inch pegs, and cut the ends flush with the braces.

Making the Tabletop and Esterilla

1 On the 22-inch plywood square , mark a circle 20 inches in diameter. Do this by hammering a tack in the center of the plywood. Tie the string to the tack, then tie the marker to the string exactly 10 inches from the tack. Holding the string taut, mark a circle on the plywood. Use the jigsaw to cut out the circle, and smooth the edges with the belt sander.

2 Take the 20-inch bamboo lengths for the esterilla, and use the splitting knife and mallet to split each in half. Take five split lengths, and use the esterilla technique described on page 49 to flatten the bamboo. Use the

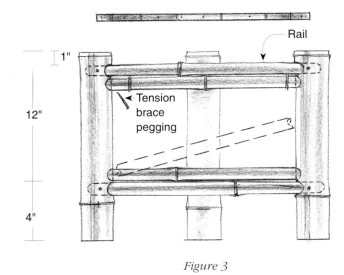

Figure 3

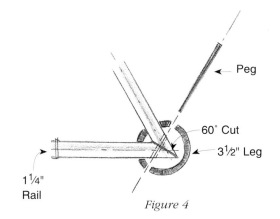

Figure 4

wood planer to shave off the rough inner wall to achieve a flat, uniform surface.

3 Refer to figure 5. Use the brush to apply a thin layer of glue to the plywood tabletop. Place the esterilla pieces on the glued surface. Start at one end of the tabletop and move across, adjusting the pieces so the nodes are lined up in the center.

4 Once all the pieces are laid out, lay a 22-inch wood strip across the tabletop center. Use the brad nails to temporarily secure the wood strips through the esterilla into the plywood, letting the nail heads stick up above

the wood just enough to be easily removed later with a hammer. As you move along, use your fingers to pull the edges of the pieces close together, making sure there are no gaps.

5 Cut the remaining wood strips into pieces, nailing the longer ones in the center of each half-circle, and shorter ones along the edges. Let dry overnight.

6 With the wood strips still attached, turn the tabletop over, and use the fine-tooth saw to cut off the esterilla that over-hangs the edge. Pull off the wood strips and nails, and remove any

glue with the wet rag.

7 Make the trim for the tabletop edge by splitting the 6-foot culm. From one half of the split culm, make a splint slightly wider than the thickness of the plywood and esterilla combined. Run the splint through the planer until it's flattened and thinned enough so it can be bent around the edge of the tabletop.

8 Now you'll install the trim; refer to figure 6. You'll need to work quickly; nearby in your work area, assemble the drill, the drill bit that's the same diameter as the furniture tacks, the hammer, and the tacks. Now, heat the strip. If

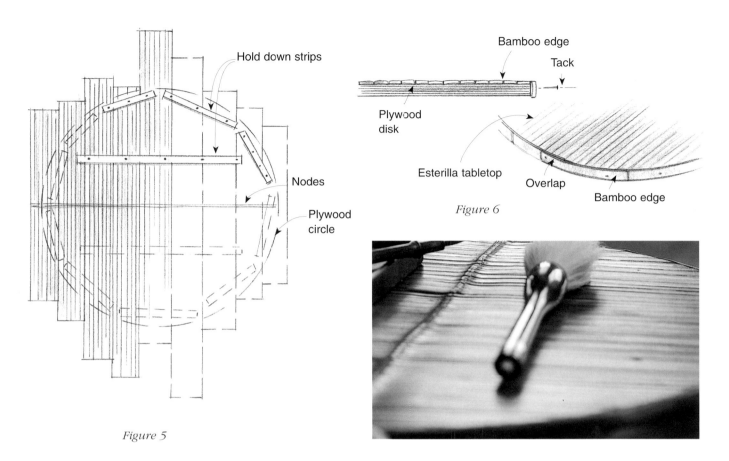

Hold down strips

Nodes

Plywood
circle

Figure 5

Bamboo edge

Tack

Plywood
disk

Esterilla tabletop

Overlap

Bamboo edge

Figure 6

you're using a propane torch, lay the strip on concrete outdoors, outer wall up, and run the flame over it in sweeping motions until it becomes loose and malleable. If you use a different flame source, hold the outer wall toward the flame.

9 Brush glue onto the inside of the trim and on the tabletop edge. Pick a starting point on the table edge. Drill a pilot hole through one end of the trim, press the trim to the edge, and hammer a tack through at the starting point. Continue attaching every 8 to 10 inches until you complete the edge. Overlap at

the end by 3 inches, apply more glue, drill another pilot hole and hammer in the last tack. Use the fine-tooth saw to cut through the strip and slightly into the underlying trim to make a flat join. Use the wet rag to remove excess glue, and let dry.

Metric Equivalents

¼"	6mm	2"	5.1 cm	14"	35.6 cm
⅜"	9.5 mm	3"	7.6 cm	16"	40.6 cm
½"	1.3 cm	3½"	8.9 cm	18"	45.7 cm
⅝"	1.6 cm	4"	10.2 cm	20"	50.8 cm
¾"	1.9 cm	6½"	16.5 cm	21"	53.3 cm
1"	2.5 cm	8"	20.3 cm	22"	55.9 cm
1⅛"	2.8 cm	10"	25.4 cm	24"	61 cm
1¼"	3.2 cm	12"	30.5 cm	36"	0.9 m
				6'	1.8 m

Side Table
of Black Bamboo and Stone

*with assistance
from Michel Spaan*

THIS ELEGANT TABLE WAS DESIGNED TO COMPLEMENT THE STONE SLAB THAT SERVES AS ITS TOP. THE STAND IS PURPOSEFULLY SIMPLE, THE RICH BLACK OF THE BLACK BAMBOO (*PHYLLOSTACHYS NIGRA*) POLES ACCENTUATING THE RIPPLING LAYERS AND SUBTLE COLORS OF THE TENNESSEE SANDSTONE. FOR YOUR OWN SIDE TABLE, PICK ANY FLAT STONE WITH SIMILAR DIMENSIONS THAT PLEASES YOU. IF YOUR STONE NEEDS TRIMMING, YOUR LOCAL STONE YARD OR STONEMASON CAN DO IT FOR YOU.

Materials

Wood stain, brownish black color

4 wooden dowels, each 36 inches long, ⅜-inch diameter, to serve as inner core of legs

2 slabs of stone, relatively flat, approximate dimensions: one slab 23½ x 15 inches, ¾ to 1¼ inch thick, to serve as top; and one slab 12½ x 10½ inches, ⅜ to ½ inch thick, to serve as shelf

Tools and Supplies

Clean rag

Fine-tooth saw

Miter box

Drill press with ½-inch Forstner bit

Rebar, 48 inches long

Hammer

Measuring tape

Round rasp

Level

Toothpicks

Sanding sticks (see page 47), with medium and coarse sandpaper

Silicone glue

Cutting List

Description	Qty.	Material	Dimensions
Legs	4	Bamboo poles	1" x 21½"
Feet	4	Bamboo poles*	1" x 4½"
Front rails	3	Bamboo poles	1" x 16"
Back rails	3	Bamboo poles	1" x 22"
Side rails	6	Bamboo poles*	1" x 13"
Sleeves	4	Scrap bamboo	¾" x 36"

*With one end cut below the node

Instructions

1 Use the rag to apply two coats of the wood stain to the bamboo legs, feet, and rails. Let dry between coats.

2 Use the miter box and fine-tooth saw to cut the ends of the rails at 45° angles. Use the Forstner bit to drill a hole through both walls of the rails, 2½ inches from each end. Before you drill, make sure all the angled ends point in the same direction and that the two holes in each length are aligned since they will be slipped over the rigid dowels.

3 Use the rebar and hammer to knock out the diaphragms of the 21½-inch legs. Use the saw, rasp, and sanding poles to make fishmouth cuts (see page 47) at both ends of the legs.

4 Use the rebar and hammer to knock out the diaphragms of the 4½-inch feet, leaving the node at the bottom intact. Ream out the node membranes. Use the saw, rasp, and sanding poles to make curved fishmouth cuts at the top of the feet.

5 Begin the stacking process by making sleeves for the feet. Insert the measuring tape into the

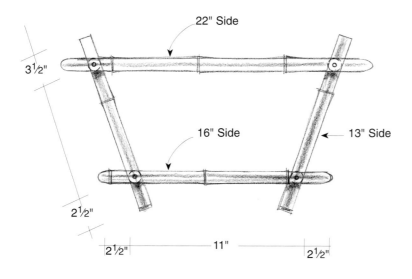

22" Side

3½"

16" Side ← 13" Side

2½"

2½" ⟷ 11" ⟷ 2½"

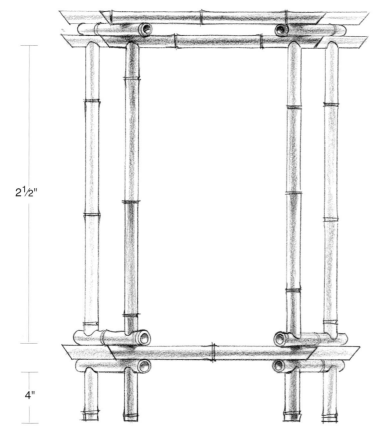

2½"

4"

hollow walls of each foot, and measure the distance from the base to the lower curved edge of the top. Transfer this measurement to a node-free length of the ¾-inch scrap bamboo and cut.

6 Place the end of a dowel in each foot. Insert the bamboo sleeve over the dowel into the foot, situating it between the dowel and the inner wall of the pole. Check for fit and make sure that the top edge of the sleeve is below the fishmouth cut. Repeat with the other three feet.

7 Slide a side rail onto two of the dowel assemblies. Use the level to determine if they're even. A slight unevenness may correct itself when other lengths are stacked, but if it is off more than ½ inch, correct it by cutting a new foot. If there are gaps where the curved top of the feet meet the horizontals, rework the fishmouth cut to get a smooth joint. Repeat with the other side rails and two feet.

8 Now remove the side rails, dowel, and sleeve and squeeze a generous amount of silicone glue into the base of the foot. Reinsert the dowel and the sleeve into the foot. At the top, apply more glue, using a toothpick to work the glue into any spaces. The glue will help keep the stand from wobbling. Repeat with the other side.

9 Join the sides together by sliding a back and front rail over the sets of dowels. The structure is now beginning to look like a four-sided table base! Use the level and adjust. Now, slide a pair of side rails through the dowels.

10 Slide the legs over the dowels, checking for a good fit at the fishmouth joint. Then slide the front and back rails onto the dowels, again checking that the fishmouth cut fits well against the bamboo. When you're satisfied that the pieces fit together and are level, remove the front and back horizontals and the legs. Repeat the technique in step 5 to make sleeves for the top and bottom ends of each leg. (The sleeves for the legs don't have to fill the entire length of the leg, they can be just long enough to stabilize the assembly.) Be sure to squeeze glue into the sleeve and dowel assemblies and use a toothpick to push it in.

11 Restack the last of the rails, checking and adjusting each one for level and fit. At this point, the stand is now assembled. Saw the dowels 1 inch above the top rails. Center the stone slab on the top of the dowels. Check with the level; if all is well, saw off the dowels flush with the bamboo, then rest the slab directly on the bamboo. If the slab is irregular, mark and cut each dowel so the stone rests solidly.

12 Place the smaller stone on the lower horizontals to create the shelf.

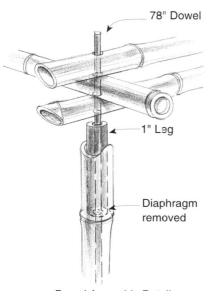

78" Dowel

1" Leg

Diaphragm removed

Dowel Assembly Detail

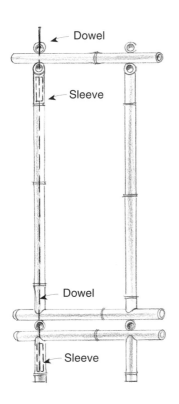

Dowel

Sleeve

Dowel

Sleeve

Metric Equivalents

⅜"	9.5 mm
½"	1.3 cm
¾"	1.9 cm
1"	2.5 cm
1¼"	3.2 cm
2½"	6.4 cm
4½"	11.4 cm
10½"	26.7 cm
12½"	31.8 cm
13"	33 cm
15"	38.1 cm
16"	40.6 cm
21½"	54.6 cm
22"	55.9 cm
23½"	59.7 cm
36"	91.4 cm
48"	121.9 cm

Elegant Vases

ITH A FEW WELL-PLACED CUTS, YOU CAN TRANSFORM CAST-OFF LENGTHS OF BAMBOO INTO SOPHISTICATED FLOWER VASES. LENGTHS TAKEN FROM THE UPPER MID-CULM MAKE LONG, SLENDER VASES, WHILE THOSE TAKEN FROM THE BASE OF THE CULM PROVIDE CLOSE-SPACED NODES FOR MULTIPLE FLOWER OPENINGS. THE INTERNODES HOLD SMALL AMOUNTS OF WATER, ENOUGH TO KEEP CUT FLOWERS FRESH FOR SEVERAL DAYS. IF YOU WISH TO HAVE A HANGING VASE, DRILL AN UPWARDLY-ANGLED HOLE BELOW THE TOP NODE IN THE BACK. HAMMER A FINISHING NAIL IN THE WALL AND HANG THE VASE.

Instructions

Simple Vase

1 Working with the 11-inch length of bamboo, make a cut straight across the base ½ inch below the node, so the vase sits upright.

2 Select and mark the side of the culm to be the front of the vase. Do this by rotating the piece, looking for interesting scarring or coloration. Place the bamboo in the miter box, and line up the top with the 45° angle slots in such a way that the elongated cut you are about to make will open at the front. Saw, being careful not to feather the outer wall.

3 Use the sanding sponge to lightly sand the edges. Brush the wood stain on the outer walls. Let dry. If desired, use the sealer to coat the inside of the culm.

Vase with Two Openings

1 At the base of the 14-inch length of bamboo, use the saw to make a cut straight across, ½ inch below the node, so the vase sits upright.

2 Choose the side of the culm to serve as the front of the vase. Place it in the miter box and line up the top of the bamboo with the 45°-angle slots in such a way that the elongated cut you are about to make will open toward the front. Use the saw to make a diagonal cut ¼ inch above the front of the node and 2 inches above the back of the node.

3 Now you'll mark the placement of the holes. Measure halfway up from the bottom node and mark a line across. Mark the same distance above the middle node. Cut straight across each line, ⅝-inch deep into the pole.

4 Mark 1½ inch above the bottom cut, and secure the bamboo firmly in place. Position the saw at the mark, angling it into the culm and down, along an imaginary line that will meet the inside ends of the cut. Saw, keeping the blade at an steady angle, until you reach the cut. Pop out the piece. Repeat to create the upper hole, this time starting the cut 2 inches above the cut.

Materials

1 length of black bamboo, 11 inches long, 2⅜ inches in diameter, with one node at the base

1 length of black bamboo 14 inches long, 2¾ inches in diameter, with three nodes

Wood stain, in brownish black color

Tools and Supplies

Fine-tooth saw

Pencil

Miter box

Sanding sponge

Paintbrush

Wood sealer (optional)

Metric Equivalents

¼"	6 mm
½"	1.3 cm
⅝"	1.6 cm
1½"	3.8 cm
2"	5.1 cm
2⅜"	6 cm
2¾"	7 cm
11"	27.9 cm
14"	35.6 cm

5 Use the sanding sponge to lightly sand the edges of the vase, and brush the stain on the outer walls. Let dry.

6 If desired, coat the inner walls with the sealer and let dry.

Triple Serving Vessel

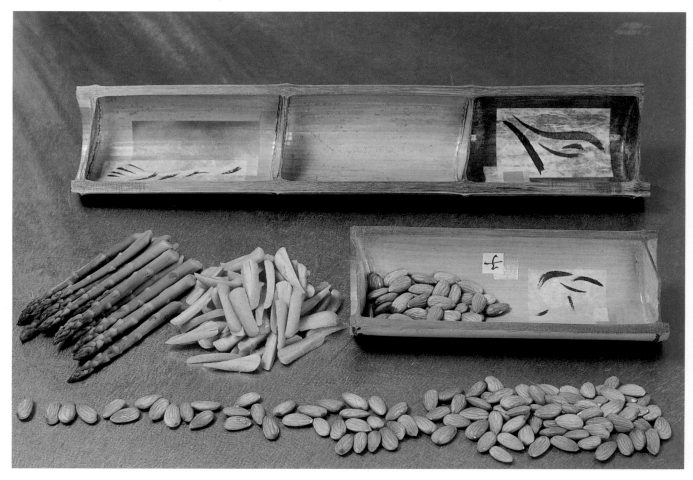

Designer, Peter Gallagher

IANT *MOSO* BAMBOO FROM JAPAN COMBINES WITH EARTH-TONED WOOD STAIN AND SILVER LEAF TO CREATE THIS SERVING VESSEL WITH THREE SPACIOUS COMPART-MENTS. FINE ASIAN PAPER IS INFUSED WITH THE BLACK INK OF *SUMI* BRUSH STROKES AND LAID WITHIN THE CURVED HOLLOWS OF THE BAMBOO. YOU'LL WANT TO PRESENT VEGETABLES, FRUITS, OR NUTS IN THIS CONTAINER. IF IT NEEDS CLEANING, HAND WASH WITH A MILD DETERGENT, AND WIPE GENTLY.

Instructions

1 Use the scrub brush and bleach solution to wash the outside of the bamboo. Dry with the cloth.

2 Use the saw to cut straight across the ends of the bamboo, close to the nodes.

3 Use the splitting knife and mallet to split the bamboo. Smooth the edges with the knife or shaver, then sand until smooth.

4 Use the scrub brush and bleach solution to clean the diaphragms and inner walls thoroughly. Let dry.

5 Use the artist's brush to apply the brownish black stain to the edges of the container, and the plum stain to one

of the inner compartments. Let dry.

6 Use the cloth measuring tape to measure the inner dimensions of the unpainted inner sections of the vessel. Use the ruler and marker to transfer the dimensions to the white paper, and use the scissors to cut out the patterns. Outline the patterns on the Japanese paper, then cut out.

7 Thin the white glue with water, then brush it onto the back of the pieces of Japanese paper. Adhere the papers to the inner walls. Let dry.

8 Use the brush to apply four coats of the varnish to seal the vessel, allowing it to dry between coats.

Bamboo vessels.
DESIGN AND PHOTO BY PETER GALLAGHER

Materials

1 length of moso bamboo (*Phyllostachys pubescens*) 24 inches long, 4½ inches in diameter, with four nodes taken from the base of the culm

Wood stains in brownish black and plum colors

Handmade Japanese paper decorated with sumi brushwork and silver leaf*

Food-safe varnish

*See page 156 for Asian paper suppliers.

Tools and Supplies

Scrub brush

Mild water and bleach solution

Cloth

Fine-tooth saw

Splitting knife

Mallet

Knife or shaver

Sanding sponge

Artist's brush

Cloth measuring tape

Ruler

Pencil

Scissors

White glue

Brush for applying glue

Small paintbrush

Metric Equivalents

4½"	11.4 cm
24"	61 cm

Sushi Trays

OUR SUSHI MAY COME FROM THE TAKE-OUT COUNTER OF A LOCAL

JAPANESE RESTAURANT, BUT YOU CAN SERVE IT ON YOUR OWN

HANDCRAFTED TRAYS OF FRESH GREEN BAMBOO. THE TRAYS ARE

A GOOD WAY TO PRACTICE SPLITTING, TOO.

Materials

3 fresh, green lengths of bamboo, each 6½ inches long, 3 inches in diameter, without nodes

Metric Equivalents

3" 7.6 cm
6½" 16.5 cm

Tools and Supplies

Water

Bleach

Scrub brush

Cloth

Measuring tape

Fine-tip permanent marker

Fine-tooth saw

Splitting knife

Mallet

Instructions

1 Use the scrub brush and a weak solution of water and bleach to wash the outside of the bamboo. Dry with the cloth.

2 Use the hammer and mallet to split each length into four pieces, making a total of 12 splints.

3 Use the knife or shaver to lightly remove the rough edges along the sides of the trays. You're ready to serve!

Tea Cups and Napkin Rings

Instructions

Making the Teacups

1 Use the scrub brush and water to clean the outside of the 4-inch pieces of bamboo. Wipe dry with the cloth.

2 Use the saw to cut each length evenly across the base, 3/8 inch below the node, so the piece sits upright.

3 Measure and mark 3½ inches up from the base of each piece.

4 Cut at the mark you made, sawing the length carefully to prevent shredding. Sand the sharp edges of the inner and outer rim until smooth.

5 Use the bottle brush and water to scrub the inside of the cups, then turn them upside down to dry.

S IMPLICITY IS THE KEY TO ARTFUL ENTERTAINING, AND THESE BAMBOO DINING UTENSILS ARE LOVELY TO LOOK AT AND USE. THEY'RE SO EASY TO MAKE, PERHAPS YOU'LL WANT TO MAKE SOME AS TAKE-HOME GIFTS.

Materials

4 fresh, green lengths of bamboo, each 4 inches long, 1¾ inch in diameter

1 fresh, green length of bamboo, 8 inches long, 1¾ inch in diameter

Tools and Supplies

Scrub brush

Cloth

Fine-tooth saw

Measuring tape

Fine-tip permanent marker

Knife

Sanding sponge

Bottle brush

Metric Equivalents

3/8"	9.5 mm
1⅝"	4.1 cm
1¾"	4.4 cm
3½"	8.9 cm
4"	10.2 cm
8"	20.3 cm

Making the Napkin Rings

1 With the scrub brush and water, wash the outside of the 8-inch length of bamboo. Dry with the cloth.

2 Measure and cut four pieces from the 8-inch length, each 1⅝ inches long. Use the sanding sponge to smooth the ends.

Shoji Window Screen

with assistance from Randall Ray and Jean Clark

THIS WINDOW TREATMENT WAS INSPIRED BY JAPANESE *SHOJI* SCREENS, SLIDING WALL

PANELS OF HEAVY PAPER FRAMED WITH WOOD. THE SPLINTS ARE CRAFTED FROM ONE

LENGTH OF BLACK BAMBOO, ADDING TO THE SYMMETRY OF THE SQUARE AND CIRCLE

SHAPES, WHILE THE PROMINENT, IRREGULAR NODES OF THE BAMBOO ADD TEXTURAL CONTRAST.

THE *KOZO* PAPER, SHOWN IN THE WINDOW SCREEN IN THE PHOTOGRAPH, IS MADE FROM MULBERRY

FIBER AND IS SIMILAR TO TRADITIONAL SHOJI PAPER.

Instructions

1 Determine the center of the circle on the plywood square. Mark the midpoint of each side at 10 inches, then mark 1½ inches in from each of the midpoints. Draw lines that connect the midpoints, intersecting in the middle. This is the center of the square. Drive a nail into the center. Tie one end of the string around the pencil, and tie the other end to the nail exactly 8½ inches from the upright pencil. Keeping the string taut, mark the circle.

2 Drill a ⅝-inch hole, or one large enough to accommodate the jigsaw blade, inside the circle near the circle's edge. Saw from the hole to the outline, approaching at an angle, then saw around the circle. Remove the circular piece of scrap. Use the sanding sponge to smooth the inner edges.

3 Brush the plywood frame and lengths of molding with two coats of the wood stain. Let dry between coats. Apply the sealer and let dry.

4 Use the saw and miter box to cut four lengths from the molding, each 21 inches long and ends mitered to a 45° angle so the pieces of molding fit together in a square. Use the finishing nails to attach the molding to the edge of the plywood, creating the outer frame.

Materials

Scrap plywood, ½" x 2C" x 20"

20 1¼-inch ringed nails

1 sheet of Japanese shoji paper, 20" x 20"

Wood stain, walnut color

Wood sealer

20 1-inch finishing nails

Double-sided tape

2 ½-inch screw eyes

2 cup hooks

Monofilament

Tools and Supplies

Measuring tape

Pencil

Hammer

15-inch piece of string

Power drill and assorted drill bits

Jigsaw

Sanding sponge

Miter box

Wood glue

Paintbrush

Nail punch

Cutting List

Description	Qty.	Material	Dimensions
Verticals	3	Black bamboo poles	¾" x 17¼"
Horizontals	1	Black bamboo pole, curved, split in half	¾" x 17¼"
Top, bottom, and center horizontals	3	Bamboo splints	1¾" x 16"
Frame base	1	Plywood	½" x 20" x 20"
Outer frame	1	Molding with rounded edges	½" x 1½" x 48"

Metric Equivalents

½"	1.3 cm	2"	5.1 cm	15"	38.1 cm
⅝"	1.6 cm	2½"	1.3 cm	15½"	39.4 cm
¾"	1.9 cm	5"	12.7 cm	16"	40.6 cm
1"	2.5 cm	8"	20.3 cm	17¼"	43.8 cm
1¼"	3.2 cm	8½"	21.6 cm	20"	50.8 cm
1½"	3.8 cm	10"	25.4 cm	21"	53.3 cm
1¾"	4.4 cm	12½"	31.8 cm	48"	1.2 m

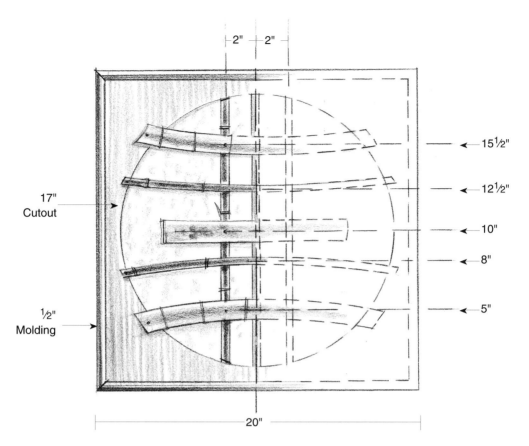

2"   2"

17"
Cutout

½"
Molding

15½"

12½"

10"

8"

5"

20"

5 On a flat work surface, place the frame on top of the 20 x 20-inch piece of scrap wood. The flat square will give height and surface to the bamboo pieces you're about to install within the circle.

6 Mark and cut the three vertical poles so one is directly in the center and the others are 2 inches on either side, cut at a slight angle to fit snugly between the opening. Turn it over so you're now working from the back of the frame. Attach the verticals by drilling a pilot hole through the ends of the bamboo at an angle into the wood. Hammer the ringed nails into the plywood and use the nail punch to drive the

head in further. Turn it over to the front.

7 Lay the horizontal pieces across the verticals. From bottom to top, measure and mark the center points at 5, 8, 10, 12½ and 15½ inches. Mark and cut the ends of the center horizontal, and cut the other ends at an angle to follow the curve.

8 Attach the middle piece to the left and right bamboo verticals. Drill a pilot hole using a bit the same size as the ringed nail. Coat a ringed nail with wood glue. Insert the nail through the bamboo wall, and gently tap the head with the nail punch and hammer.

9 Attach the other bamboo pieces. Drill pilot holes in the corners, and hammer in ringed nails, followed by a tap with the hammer and nail punch. No need to add glue, as the nails firmly grab the plywood.

10 Affix the double-sided tape to the back of the frame, ¾ inch from the inside of the molding. Carefully center the paper and lay the edges over the tape, pressing down along all four sides to secure.

11 Attach the screw eyes to the top of the frame 2 inches from each side. Tie the monofilament to the screw eyes, and hang from the cup hooks at the desired location.

Bamboo Doormat

Designers, Yucatan Bamboo

THIS HANDSOME MAT OF IRON BAMBOO (*DENDROCALAMUS STRICTUS*) WORKS WELL AT THE EDGE OF A PATIO, PORCH, OR POOL. ORIGINALLY DESIGNED FOR THE STONE COURTYARDS OF A MEXICAN HACIENDA-TURNED-HOTEL, THIS RESILIENT, TIGHTLY STRUNG MAT KEEPS FEET ABOVE GROUND LEVEL AND ALLOWS WATER TO DRAIN THROUGH. A DRILL PRESS IS NEEDED FOR THIS PROJECT; IF YOU DON'T HAVE ONE, TAKE THE PIECES TO A LOCAL WOOD SHOP. THE PROCESS FOR MAKING THIS DOORMAT IS FULLY PATENTED, BUT INDIVIDUALS ARE WELCOME TO MAKE MATS FOR THEIR PERSONAL USE.

Instructions

1 Set up the jig on the drill press. Place the 36-inch plank on the surface of the press, up against a straight edge. Make sure the plank is level. Lower the drill press with the ⅜-inch bit until it makes an indentation in the plank. Measure 3 inches beyond this point, and attach with screws the 6-inch strip of wood to serve as the stop. Make sure the stop is square at a 90° angle to the plank. Line the lengths of bamboo up against the stop to ensure all drilled holes will be a uniform 3 inches from the end.

2 Examine each pole to determine where the hole should be drilled. If the pole has a slight curve, rotate it so the inside of the curve faces upward. Place one end up against the stop with the section to be drilled lying flat against the plank under the drill press. It's acceptable for part of the pole to curve slightly upward off the plank. Use the drill press to make the hole, repeating the process (at one end only) with all of the lengths.

3 Measure 27 inches from the stop, exactly in line with the drill press indentation, and mark. Drill a ⅛-inch pilot hole at the mark and through the plank at a 90° angle. Turn over the plank and drive the 3-inch nail through it. Return the plank to the drill press, the tip of the nail pointed up.

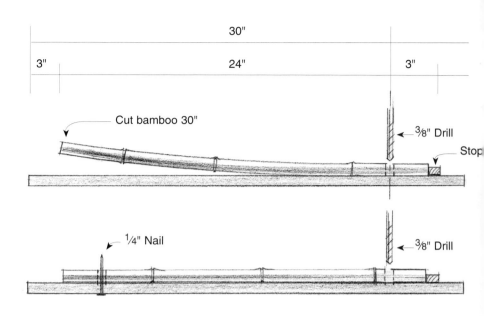

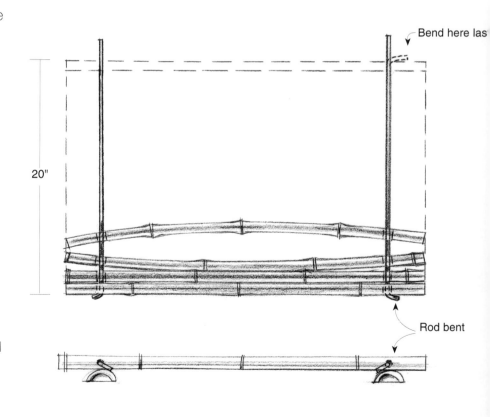

Materials

1 x 6 wood plank, 36" long

1 strip of wood, ½" x 1" x 6"

5 drywall screws, 1⅝ inches long

3-inch nail, ⅜-inch diameter

2 lengths of ¼-inch galvanized
steel rod, each 24 inches long

String

Wood sealer

Tools and Supplies

T-square

Drill press with ⅜-inch drill bit

Level

Measuring tape or ruler

Hammer

Vise

Pliers

Hacksaw

Splitting knife

Mallet

Power drill with assorted drill bits

Paintbrush

Cutting List

Description	Qty.	Material	Dimensions
Mat surface	20	Iron bamboo pole	¾" to 1" x 30"
Feet	1	Iron bamboo pole*	2" to 2½" x 22"

*It's important to select a very straight length of bamboo.

Metric Equivalents

⅛"	3 mm	3"	7.6 cm
¼"	6 mm	6"	15.2 cm
⅜"	9.5 mm	22"	55.9 cm
½"	1.3 cm	24"	61 cm
¾"	1.9 cm	27"	68.6 cm
1"	2.5 cm	30"	81.3 cm
1⅝"	4.1 cm	36"	0.9 m
2"	5.1 cm		
2½"	6.4 cm		

4 Slip the drilled end of a pole over the nail, pressing the other end against the stop. Drill a hole through the bamboo. Repeat with all the poles.

5 Use the vise and pliers to make a ½-inch, 90° bend at the end of each steel rod.

6 Working with one pole at a time, determine its curvature for placement on the rods. String the two rods through the holes in the bamboo, alternating the slight curve facing up with the slight curve facing down, to create a flat surface. Use the mallet to hammer each length close to the next. String all the poles.

7 Use the pliers and hammer to bend the wire rods at a right angle, securing the poles together. Use a hacksaw to cut off the excess.

8 Now you'll attach the feet. Use the knife and mallet to split the 22-inch length. Line up the halves along the wire rods, the curved side up against the bamboo, the flat side resting evenly on the floor.

9 Use the power drill and screws to attach the feet to the bottom of the mat, driving them in from the under side of the mat.

10 Brush the wood sealer on all surfaces of the mat and let dry.

Chopsticks & Chopsticks Rests

OU CAN PRACTICE SPLITTING BAMBOO BY CRAFTING YOUR OWN CHOPSTICKS AND CLEVER RESTS TO SET THEM ON. TIE TOGETHER A BUNDLE WITH RIBBON OR RAFFIA, AND YOU HAVE A CHARMING GIFT THE NEXT TIME YOU DINE AT A FRIEND'S HOUSE.

Instructions

Making the Chopsticks

1 Use the scrub brush and a weak solution of water and bleach to wash the outside of the bamboo. Dry with the cloth.

2 Use the knife and mallet to split the length in half. Reserve half to make the chopstick rests, and split the other half into quarters.

3 Take one of the four splints you just made. Use the knife or shaver and sanding sponge to smooth the rough edges and the inside wall until the bamboo feels good in your hand.

4 Use the knife to shave off the outer corners and form a blunt-ended point on one end of the chopsticks. This end will serve to pick up food. Sand the edges.

5 Hold the splint upright with its node end down. Place the splitting knife exactly in the center of the blunt-pointed end. Tap the knife with the mallet. Carefully twist and move the knife to split to, but not through, the node. The person who uses the chopsticks will pull them apart.

6 Repeat steps 3 through 5 with the remaining three splints.

Chopstick Rests

1 Take the split half you put aside in step 2 above. Use the saw to cut it into four pieces, each ¾ inch wide. Lay the resulting half-circle pieces of bamboo on their sides.

2 With the splitting knife and mallet, remove ½ inch from the ends of the half circles, so that when you turn them upright to hold the chopsticks, their height is ¾ inch.

Materials

1 fresh, green length of bamboo, 10¼ inches long, 2¼ inches in diameter, with one node 1½ inches from the base

Tools and Supplies

Water
Bleach
Scrub brush
Cloth
Splitting knife
Mallet
Knife or shaver
Sanding sponge

Metric Equivalents

½"	1.3 cm
¾"	1.9 cm
1½"	3.8 cm
2¼"	5.7 cm
10¼"	26 cm

Formal Japanese Flower Container

Designer, Brother Konomo Utsumi

THIS ELEGANT CONTAINER FOCUSES ON THE SIMPLICITY, FORM, AND UTILITY OF BAMBOO. WHEN YOU MAKE THIS FLOWER CONTAINER, YOU'LL BE CREATING AN OBJECT THAT REFLECTS CONTEMPORARY JAPANESE AESTHETICS.

Instructions

1 Sweat the bamboo poles, using the process explained on page 50.

2 Cut a 7-inch length from one of the poles. Use the knife and mallet to split it, creating two feet.

3 Arrange the three poles side by side. Position the center area of each length, where the openings will be cut, between nodes. Mark the center points.

4 Keeping the centers lined up and measuring from the left side, mark one length at 35 inches, one length at 21 inches, and one length at 27 inches. Use the saw to cut the lengths straight across at the nodes. Cut the other ends on an angle to create an open elongated oval.

5 Now you'll make the center openings. Starting with the 35-inch length, measure 14 inches from the left end, and use the saw to cut straight down about half way through the culm. From that cut, measure up 1¾ inch and saw on an angle into the culm to meet the cut. Pop out the inner piece. Repeat this process with the other two lengths, cutting the 21¾-inch length 9 inches up from the base, and the 27¼-inch piece 12 inches up from the base.

6 Use the knife to slice off the black outer layer of the bam-

Materials

3 poles of black bamboo
(*Phyllostachys nigra*) each 36 inches long, 1 inch in diameter

Metric Equivalents

1"	2.5 cm
1½"	3.8 cm
1¾"	4.4 cm
5"	12.7 cm
7"	17.8 cm
9"	22.9 cm
12"	30.5 cm
14"	35.6 cm
21"	53.34 cm
21¾"	55.24 cm
27"	68.6 cm

Tools and Supplies

Fine-tooth saw

Splitting knife

Mallet

Fine-tip permanent marker

Measuring tape

Knife

Power drill and assorted drill bits

6 drywall screws, 1 to 1½ inches long

Water sealer

Paintbrush

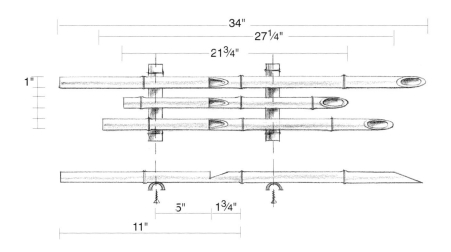

boo and to make decorative stripes on the ends, at the center openings, and on the feet.

7 Attach the feet by centering the three poles with their openings facedown. Measure and mark 5 inches below the straight cut of the opening and 7 inches above it.

8 Center the feet, curved sides facing down, along the points you marked. Drill pilot holes, and drive the 1-inch screws through the feet and into the poles.

9 Use the brush to apply the sealer on the outer surfaces and inside the center and end openings of the vase. Let dry.

Chapter 8

Fences and Screens

YOU CAN USE BAMBOO TO BUILD FENCES AND SCREENS IN A WIDE RANGE OF STYLES TO COMPLE-MENT GARDENS, PONDS, COURTYARDS, DECKS, AND PATIOS. A LOW, SEE-THROUGH FENCE CAN EASE THE TRANSITION BETWEEN DIVERSE AREAS IN A LAND-SCAPE, WHILE A TALL, NAR-ROW SCREEN ADJACENT TO A HOUSE CAN SUGGEST A SENSE OF PRIVACY. SOLID WALLS OF BAMBOO CAN CONVENIENTLY SHUT OUT AN UNDESIRABLE VIEW, WHILE AN OPEN BAMBOO FENCE CAN EMBRACE THE LANDSCAPE BEYOND. ONCE YOU UNDERSTAND THE BASIC CONSTRUCTION PROCESSES FOR MAKING FENCES AND SCREENS, YOU'LL BE ABLE TO CREATE VARIA-TIONS THAT SUIT YOUR LANDSCAPE AND REFLECT YOUR PERSONAL STYLE.

Fence and Screen Construction

Fences and screens consist of five main elements as shown on page 102. *Posts* are placed upright in the ground, joined together by *stringers*. *Verticals* are bamboo poles secured onto the stringers, and *horizontals* are decorative split lengths that attach to the verticals and stringers. *Ties* lash over the horizontals to secure them to the poles and stringers.

DESIGN AND LOCATION

Planning is the key to a successful fence or screen. Measure the intended site, and use stakes and flagging to mark its boundaries. Draw the design on graph paper, drawing to scale the length and height of the structure and the placement of stringers and verticals. Based on your own plan or any modifications you make in the following projects, develop a list of materials and supplies. Where do you build? Multi-paneled fences, as described in this book, are built on-site. Single panel screens can be constructed off-site and then installed.

POSTS

Spaced 5 to 8 feet (1.5 to 2.4 m) apart, posts act as anchors for the entire structure. Fence posts and timbers are available from home improvement and farm supply stores; buy posts that have been pressure-treated with chemicals to withstand the weather and insects. If the post is untreated, brush a wood preservative over the part of the post to be sunk in the ground. If you plan to use bamboo posts, choose stout, strong poles. When calculating length, add a minimum of 18 inches (45.7 cm) to set in the ground.

NOTCHES

Measure and mark the center placement of the stringers on the post, i.e., the point at which the middle of each stringer will join the post. From that point, draw parallel lines to mark the width of the stringer. Saw along these lines to a depth of ¾ inch (1.9 cm) or more, then use a hammer and chisel to knock out the wood in between. At this point, stain the posts if desired and let dry.

SETTING POSTS

Using a post hole digger, dig holes to a depth of 18 to 24 inches (45.7 to 61 cm). Be mindful of the possibility of underground electrical and television cables, irrigation hoses, and water and gas lines. Place 2 inches (5 cm) of crushed rock into the hole, and tamp. Place the posts in the holes and fill in

Split bamboo creates diamond patterns over panels of reed fencing.
DESIGN AND PHOTO BY ROBERT SMALL

Bamboo poles are attached on the diagonal to create this fence at the Bamboo Farm, Savannah, Georgia.

In this bamboo fence, "windows" open to the landscape beyond.
DESIGN AND PHOTO BY ROBERT SMALL

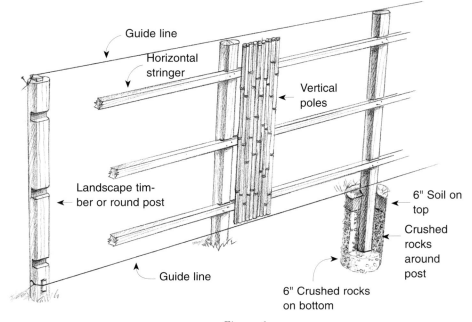

Guide line

Horizontal stringer

Vertical poles

Landscape timber or round post

6" Soil on top

Crushed rocks around post

Guide line

6" Crushed rocks on bottom

Figure 1

6 inches (15.2 cm) with the crushed rock. Use a piece of rebar to jiggle and tamp the rocks to remove air pockets to stabilize the pole. Use a level to check that the post stands straight, and tamp the rock again. Pour concrete or add gravel, and check the post again with the level. Continue to add gravel or concrete. Just below ground level, fill with soil.

STRINGERS

Pressure-treated 2 x 2 lumber makes good, all-purpose stringers for bamboo fences and screens. It's relatively light, easy to work with, and gives a flat surface on which to attach the verticals. Bamboo can also be used as stringers, but be sure to choose straight, even poles. After measuring and cutting the stringers, stain them (if desired) and let dry. To attach stringers to posts, position them and screw into place.

VERTICALS

Filling in the framework with vertical bamboo poles is the fun part! Verticals should not rest directly on the ground or decay will set in. It's best to secure verticals several inches above ground level or rest them on a stone wall. Likewise, for outdoor projects, it's not a good idea to let rain collect in bamboo's upright hollow tubes. For this reason, when you're cutting vertical poles, plan so you cut just above the top node. That way what little water collects in the concave node membrane will evaporate quickly.

You can install a guide string to keep the top and bottom of the vertical bamboo poles in line.

Mark the posts with the position, and drill a 1-inch (2.5 cm) screw into each mark. Stretch mason's cord from screw to screw. Use this as well as the T-square to keep the poles even.

When working with long or heavy vertical poles, you can add a temporary ledge of 1 x 4 lumber for the poles to rest upon as you attach them to the stringers. Drive in screws at an angle through the ends of the board between the posts. After you've attached the verticals, remove the screws and ledge.

ATTACHING VERTICAL POLES

Work on one panel of your fence or screen at a time. Position a pole next to one post, and attach it by drilling a pilot hole through it and the stringer, and then nailing or screwing it into place. For the next vertical, try out different poles, rotating, and reversing tops and bottoms to find the best fit. When you find the pole and placement that feels right, clamp or hold in place, drill pilot holes, and use screws or nails to attach. Work back and forth from post to post, attaching verticals in groups until the poles meet in the middle. Choose the last group carefully to make sure there are no major gaps.

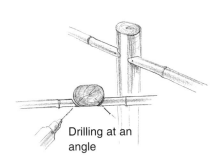

Drilling at an angle

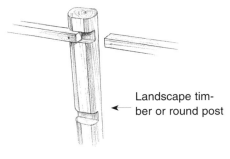

Landscape timber or round post

Figure 2

HORIZONTALS

Made of split bamboo, horizontals visually unify fences or screens, and hide screw or nail heads. To make a horizontal, choose a straight, handsome pole 2 to 3 inches (5.1 to 7.6 cm) in diameter. Carefully split it in half, to give you two lengths. Position the split bamboo horizontals on top of the verticals, which are positioned in turn over the stringers. Measure, mark, and cut the ends of the horizontals so they fit close against the post. Work with an assistant to secure them temporarily with twine to the verticals and stringers. Once they're positioned to your satisfaction, screw or tie into place.

TIES

Decorative Japanese ties are used minimally, as accents. Three to five ties per horizontal is typical, with placement often staggered. For your design, try out different arrangements on paper, then transfer the positions onto the horizontals with removable tape. Stand back and see if the pattern is too busy or too sparse, adjusting until you get a pleasing arrangement. Avoid an ambitious array until you feel comfortable making the knots. To prevent the twine from roughening and coloring your fingers, wear latex or tight-fitting work gloves.

If the poles are close together, ties are difficult to make since one pair of arms can't reach over and around the poles. It's a good idea to enlist the aid of a friend who can stand behind the fence, pull the cordage through to the back, manipulate it, and pass it back to you. If you have difficulty getting the twine through the gaps between the poles, wrap the ends with masking tape. This forms a needle-like cap which you will cut and discard when the tying is complete.

Tools and Supplies

To lay out the area and boundaries: measuring tape, stakes, flagging, mason's cord

To cut posts and stringers: power saw

To make notches in the posts: hand saw, hammer, and chisel

To dig and set posts: shovel, post-hole digger, tamper, 4-foot (1.2 m) length of rebar, crushed rock or gravel, quick-set concrete, water, and bucket or other container in which to mix the concrete

To keep fence components level: level, T-square

To attach stringers to posts and vertical poles to stringers: quick-release clamps, power drill, assorted drill bits, nails, and screws

To color and/or seal the posts: stain, paint, or sealer, paintbrush

Kenninji Fence

with assistance from Allen Fowler

HANDSOME IN ALL ITS VARIA-
TIONS, THE *KENNINJI-GAKI* IS
THE MOST COMMON SCREENING
FENCE IN JAPAN. MADE WITH WHOLE OR SPLIT
BAMBOO, IT HAS THREE TO FIVE SPLIT HORIZON-
TALS AND USES DARKENED POSTS AND TIES
FOR CONTRAST. THE CLOSE PLACEMENT OF ITS
POLES CREATES AN ATTRACTIVE PRIVACY WALL
AS WELL. THE RIGHT-ANGLED, THREE-PANELLED
FENCE SHOWN HERE HIDES TWO LARGE AIR-
CONDITIONING UNITS. HARVESTED FROM THE
OWNER'S ADJACENT BAMBOO GROVE, THE
POLES COMPLEMENT THE STONE WALKWAY AND
EARTHEN WALLS, AND PROVIDE A WARM AND
SOOTHING ENTRANCE.

Materials

Wood stain, deep brown

16 2¼-inch galvanized decking screws

136 1½-inch galvanized decking screws

12 3-inch galvanized decking screws

Double-strand, black hemp twine

Tools and Supplies

Fence-building tools and supplies listed on page 103

Fine-tooth saw

T-square

Scissors

Binder twine

Removable tape

Instructions

1 You can adapt these instructions to make as many panels as you require. First, mark the posts where the stringers will be attached. Measuring from the top of each post, mark at 2, 16, 28, and 42 inches. Make two more marks, one at 54 inches and the other at 58 inches. The first mark indicates the bottom of the vertical poles, and the second mark is the ground line. The remaining 18

Metric Equivalents

1"	2.5 cm	28"	71.1 cm
1½"	3.8 cm	42"	106.7 cm
2"	5.1 cm	54"	137 cm
2¼"	5.7 cm	57"	145 cm
2½"	6.4 cm	58"	147.4 cm
3"	7.6 cm	60"	152.4 cm
3½"	8.9 cm	63"	160 cm
16"	40.6 cm	64"	163 cm
18"	45.7 cm	76"	100.6 cm

inches of the post will be sunk into the ground.

2 Brush the posts with the dark brown stain and let dry.

3 Measure and mark notches in the posts to accommodate the stringers. Saw to a 1-inch depth at the notch lines, and use the hammer and chisel to remove the wood in between.

4 Dig holes for the posts, and set the posts in place with gravel.

Cutting List

Note: Quantities and instructions are given for 1 fence panel 64 inches long and 57 inches high.

Description	Qty.	Material	Dimensions
Posts	2	Pressure-treated landscape timbers	2½" x 3½" x 76"
Stringers	4	Pressure-treated 2 x 2 lumber	63" long
Verticals	36	Bamboo poles	1½" x 54", one end cut just above the node
Horizontals	5	Split bamboo lengths and cap	2" x 60"
Top of cap	1	Bamboo pole	1½" x 60"

5 Place the stringers into the notches on the posts. Secure with the 2¼-inch screws.

6 Attach the vertical poles to the stringers with the 1½-inch screws. Use the T-square to keep the poles straight. Beware of placing the poles too close together; you'll need gaps wide enough for several strands of twine to pass through when you make the ties.

7 Use the binder twine to temporarily attach the split horizontals over the poles to cover the screw heads. Use the 3-inch screws to secure to the stringer, and remove the twine.

8 Now add the caps to the panel. Rest one split length, open side down, on the tops of the poles. Place the other split length beside it, resting on the stringer. The caps should slightly overlap the edges. Drill pilot holes, then use the 3-inch screws to secure the split bamboo horizontals, poles, and stringers. Position the whole pole in the crevice between the split lengths. Using the 3-inch screws, secure to the top stringer.

9 Use the double-strand black twine to tie traditional Japanese ties (see page 45), covering the screw heads on the split horizontals. Along the cap, use more twine and make a double overhand knot at the ends.

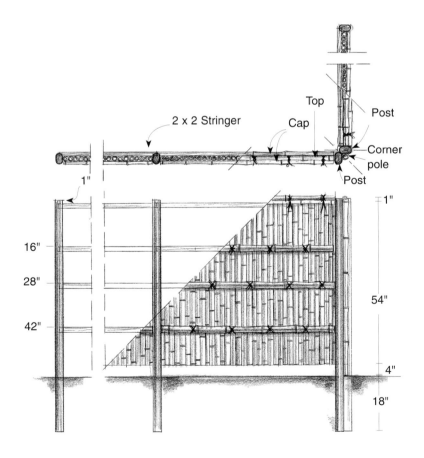

Caps ½" top

2 x 2

Horizontal 2" cap

2 x 2

2 x 2 Stringer

Cap

Top

Post

Corner pole

Post

1"

16"

28"

42"

1"

54"

4"

18"

Teppo Screens on Stone Ledges

Designer, Michel Spaan

THESE MAGNIFI-CENT SCREENS CREATE AN ENCLOSURE THAT SERENELY EMBRACES THE QUIETLY FLOW-ING WATER OF THE FOUNTAIN. THE LEFT AND RIGHT SCREENS DIFFER IN HEIGHT BECAUSE OF THE SLOPING TERRAIN, BUT THEIR TOPS ARE LEVEL WITH EACH OTHER. (THESE INSTRUCTIONS ARE ADJUSTED TO MAKE SCREENS OF EQUAL HEIGHT.) ATTACHMENTS TO UPPER TREE BRANCHES GIVE THE LARGE BACK SCREEN EXTRA STABILITY AGAINST THE WIND.

Materials

12 lengths of rebar, each 6 feet long,
4 per screen (2 for post supports
and 2 to stabilize the rear)

1,000 pounds of fieldstone pavers

15 rolls of 20-gauge copper wire

Tools and Supplies

Fence-building tools and supplies listed on
page 103

Fine-tooth saw

Wire cutters

Sledgehammer

Cutting List

Description	Qty.	Material	Dimensions
Back screen posts	2	Bamboo poles	2½" x 90"
Left and right screen posts	4	Bamboo poles	2½" x 72"
Back screen stringers	2	Bamboo poles	2½" x 11'4"
Back screen stringers	2	Bamboo poles	1½" x 10'
Back screen stringer	1	Bamboo pole	1½" x 9'9"
Left and right screen stringers	4	Bamboo poles	2½" x 6'8"
Left and right screen stringers	4	Bamboo poles	1½" x 6'
Left and right screen stringers	2	Bamboo poles	1½" x 5½'
Back screen verticals	76	Bamboo poles, varying lengths	1½" x 76" to 88", one end cut just above the node
Left and right screen verticals	64	Bamboo poles	1½" x 72", one end cut just above the node
Sleeves for rebar supports	6	Bamboo poles	1½" x 4½'

Metric Equivalents

1"	2.5 cm	72"	1.6 m	100"	2.5 m	6'8"	2 m
1½"	3.8 cm	76"	1.9m	2'	61 cm	9'9"	2.9 m
2½"	6.4 cm	80"	2 m	4'	1.2 m	10'	3 m
5"	12.7 cm	82"	2 m	4½'	1.35 m	11'4"	3.4 m
12"	30.5 cm	88"	2.2 m	5½'	1.65 m	1,000 lbs. 454 kg	
52"	132.1 cm	90"	2.25 m	6'	1.8 m		

Instructions

1 Measure the site, and use the stakes and flagging to mark the placement of the stone ledges.

2 Within each of the three ledge sites, measure the placement of the rebar, which will support the posts. Center the rebar in the areas, 100 inches apart for the back screen and 52 inches apart for the left and right screens. Use the sledgehammer to drive two lengths of the rebar 2 feet into the ground.

3 Build the stone ledges around the rebar. Dry stack the pavers within the area, fitting and adjusting to set the stones securely and achieve an even outside edge.

4 Use the hammer and rebar to knock out the diaphragms of the lower 4 feet of each bamboo post.

5 On the front of the posts, measure and mark the points where the top and bottom

stringers (not the middle stringers) will be attached, 12 inches from the top and 5 inches up from the bottom. Cut the lines of the notches on the front of the poles. Saw to a depth of 1 inch, and use the hammer and chisel to knock out the in-between piece of bamboo. Slide the posts over the rebar.

6 Now attach the top and bottom stringers to the posts by lashing with copper wire.

7 Tightly lash the unnotched middle stringers to the backsides of the posts.

8 Now attach the vertical poles to the stringers, placing the closed-node ends up and using the level to check for straightness. Note that the front poles are wedged between the stringers. All poles rest on the stone ledge.

9 Lash the poles by using one continuous roll of wire per stringer. Do this by wrapping each intersection with a basic cross tie (see page 44). Then, rather than cutting the wire when the lashing is complete, leave it intact to wrap the next pole. For the back screen, the grouping of poles is: five in front, six in back, five in front, six in back, five in front, eight in back, six in the center front, eight in back, five in front, six in back, five in front, six in back, and five in front. For the left and right screens, the grouping is: five in front, eight in back, six in the center front, eight in back, and five in front.

10 When placing the vertical poles in the back screen, use poles between 82 and 90 inches in length for the first group of poles on either side, then use poles 76 to 80 inches long for the central group. This will give a slight, inward dip to the staggered top edge.

11 Cut the ends of the stringers at an angle as shown.

12 Stabilize the screens in the back by angling two lengths of rebar between the screen and the ground. To hide the rebar and make the supports visible for safety, make bamboo "sleeves" by using a hammer and rebar to knock out the diaphragms of the 4½-foot poles. Slide the sleeves over the rebar. Hammer one end of the rebar into the ground, and wire the other end to the screen.

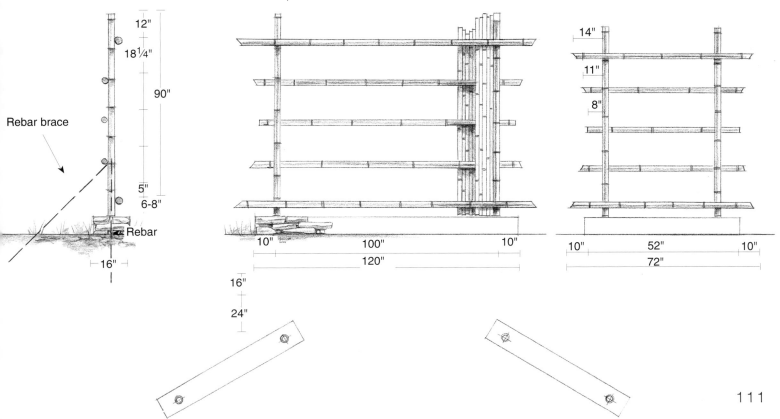

Yotsume Fence

{ IMPLE IN CONSTRUCTION, THE *YOTSUME-GAKI* IS THE MOST COMMON BAMBOO FENCE IN JAPAN. ITS POLES GIVE SUPPORT TO LEANING SHRUBBERY, AND ITS OPEN SPACES GIVE PLANTS ROOM TO SPREAD. VARIATIONS ARE MANY. YOU CAN STAGGER THE TOPS FOR A COTTAGE FEEL, USE PAIRS RATHER THAN SINGLE POLES AS VERTICALS, OR USE TALLER POSTS AND STRINGERS FOR A HIGHER FENCE. THE INSTRUCTIONS BELOW ARE FOR A 12-FOOT SECTION.

Materials

10 1⅝-inch drywall screws

50 1½-inch galvanized decking screws

Black hemp twine

Tools and Supplies

Wood stain, deep brown

Paintbrush

Fence-building tools and supplies listed on page 103

Rebar, 4 feet long or 2 feet long

Scissors

Cutting List

Description	Qty.	Material	Dimensions
Posts	2	Pressure-treated rounds	3" x 42"
Stringers	2	Bamboo poles	1½" x 12'4"
Verticals	25	Bamboo poles	1¼" to 1½" x 30", one end cut just above the node

Metric Equivalents

½"	1.3 cm	3"	7.6 cm	20"	50.8 cm	6'	1.8 m
1¼"	3.2 cm	5"	12.7 cm	30"	76.2 cm	12'	3.6 m
1½"	3.8 cm	6"	15.2 cm	42"	106.7 cm	12'4"	3.7 m
1⅝"	4.1 cm	12"	30.5 cm	4'	1.2 m		

Instructions

1 Refer to figure 1 on page 123. Measure from the top of each post and mark the center point of the stringers at 4, 28, and 48 inches. Mark the ledge placement at 56 inches and the ground line at 60 inches.

2 Cut notches for the stringers by sawing 1 inch deep into the wood, and using the hammer and chisel to knock out the in-between wood.

3 Dig post holes and set the posts 18 inches into the ground.

4 Position the ledge between the posts, and use the level to make sure it's even. Drill the 1½-inch screws into the ends at an angle to secure the ledge into the posts.

5 Use the 1½-inch screws to attach two of the 1 x 2 lengths to the front and back of the ledge. These "lips" will keep

the slender poles from slipping off the ledge.

6 Prepare the culms by shaking off the loose leaves. Use the garden pruners to cut off any leaves remaining on the branches; if left on, they will remain in the brush, darkening with mold and mildew.

7 Use the rope to make two parallel lines 7 feet apart on a large, flat surface. Lay out the

Materials

9 1½-inch galvanized decking screws

Binder twine

Black hemp twine

14 2-inch galvanized decking screws

12 2½-inch galvanized decking screws

21 1¼-inch ringed nails

Tools and Supplies

Fence-building tools and supplies listed on page 103

Garden pruners

Rope

Loppers

Scissors

Hedge clippers

Cutting List

Description	Quantity	Material	Dimensions
Posts	2	Pressure-treated 2½ x 3½ landscape timbers	6½'
Stringers	3	Pressure-treated 2 x 2 lumber	5½'
Ledge	1	Pressure-treated 1 x 4 lumber	5'
Front stringers and ledge lips	5	Pressure-treated 1 x 2 lumber	5'
Verticals	3 pick-up truckloads	Bamboo culms and branches	½" x 7'
Horizontals	2 split lengths	Bamboo poles	3" x 5' long

Metric Equivalents

½"	1.3 cm	3"	7.6 cm	56"	142.2 cm
1"	2.5 cm	3½"	8.9 cm	60"	152.4 cm
1¼"	3.2 cm	4"	10.2 cm	5'	1.5 m
1½"	3.8 cm	18"	45.7 cm	5½'	1.65 m
2"	5.1 cm	28"	71.1 cm	6½'	1.95 m
2½"	6.4 cm	48"	121.9 cm	7'	2m

bamboo top to bottom, including branches, between the two lines. Use the loppers to cut off the culms that extend beyond the bottom line. These can be used in the bundles.

8 Assemble the trimmed poles into 15 piles, which when tied, will become the verticals. Since the brush makes the upper section wider, add the poles you cut in step 7 to make the lower sec-

tions wider. Use the binder twine to wrap them into tight bundles, each about 4 inches in diameter.

9 Attach the verticals to the stringers by positioning the base of a bundle on the ledge and wedging it between the lips. With more binder twine, tie the bundles temporarily to the stringers. Work back and forth from post to post, fitting one bundle close to the next, until they meet in the middle.

10 Now you'll attach the front stringers to hold in the brush. With a friend's help, use doubled binder twine to tie together the front stringer, the verticals, and the back stringer. Pull tight! The pole and brush verticals should be sandwiched securely between the stringers. Use the 2½-inch screws, driven in at an angle, to attach the front stringers to the posts.